Military Organizations,
Complex Machines

Cornell Studies in Security Affairs

edited by Robert J. Art *and* Robert Jervis

Military Organizations, Complex Machines

MODERNIZATION IN THE
U.S. ARMED SERVICES

CHRIS C. DEMCHAK

Cornell University Press

ITHACA AND LONDON

First published 1991 by Cornell University Press.

International Standard Book Number 0-8014-2468-2
Library of Congress Catalog Card Number 90-55731
Printed in the United States of America
Librarians: Library of Congress cataloging information
appears on the last page of the book.
⊗ The paper in this book meets the minimum requirements
of the American National Standard for Information Sciences—
Permanence of Paper for Printed Library Materials, ANSI Z39.48-1984.

Contents

[v]

Preface

A paradox surrounds our complex technologies. They offer great capabilities, but they also produce massive headaches when a system fails. For example, on at least two occasions unexpected problems created by coffeepots and frozen water tanks have nearly caused large commercial airliners to crash. The loss of insulation on a small bundle of wires in a redundant oxygen tank nearly killed the astronauts of *Apollo 13*. A nest of field mice, a mislaid wrench, and stuck water valves have caused shutdowns or major accidents in nuclear power generating plants.[1] Given the complexity of such systems, these chains of events were inconceivable before the accidents.

Military forces that acquire complex technologies face this paradox multiplied many times. Critical systems on which so much depend are defeated by conditions common in wartime. In the early 1980s, a U.S. Army prototype air defense artillery gun, DIVAD, inexplicably focused on an outhouse fan rather than on its real objective and consistently missed easy targets, such as hovering helicopters (Easterbrook 1984:15). During the Falklands War, the Sea Wolf missile defense system unexpectedly refused to fire when faced with more than one attacker. And despite the military's confidence in the Rapier air defense system, the system failed

---

[1]These examples come from Charles Perrow's *Normal Accidents* (1985), one of the best treatments of these "system accidents."

when broken alloy pins caused the projectiles to drop harmlessly to the ground (Hastings and Jenkins 1983:159, 211).

Is there something inherently problematic about combining complex systems with some contexts? A prominent military historian suggests that both war and technology have their own distinct and conflicting logics. The pursuit of one inhibits the successful pursuit of the other (Van Creveld 1989:319). In the highly uncertain conditions the military faces, technologies often do not function as planned. An equally prominent sociologist believes that combining crises with technological complexity requires simultaneous centralization and decentralization in an organization, and encourages accidents (Perrow 1984).

This book is about the consequences that relying on complex machinery has for organizations like the military and offers a cautionary tale about understanding the full costs of complex technologies without rejecting the machines themselves. It looks at how the introduction of more complex equipment can increase the complexity of an organization and lead to poorer performance. The insights here are drawn from a variety of disciplines and are applicable to a variety of organizations. Students of military history will not be surprised to find that on the battlefield the more complex organizations have greater difficulty than simpler organizations. But how that complexity produces the difficulties is neither obvious nor handily demonstrated in the current literature.

This book also shows why technological complexity is important under some conditions and for some organizations, but not for others. The defense community engages intermittently in heated debate over simplicity versus complexity as organizing principles. And neither side wins, because each is certain that its preferred principle is applicable in all cases. Yet military insiders know that some complex organizations and machines work more or less well while others do not. Neither side in the debate can explain why the other can come up with examples of successes too. Similarly, students of bureaucracies often point out the "red tape" or the slowness of multiple-level bureaucracies to make decisions. But what is not sufficiently clear is why the structures behind these leisurely processes are fatally flawed for some organizations but merely irritants for others.

In an effort to explain—not bewail—the potential dark side of complexity, what I offer here is an explanation for complexity and an analytical method for viewing complexity, as it is found both in machines and in organizations. The drawbacks of the unknowns within a complex system are present in both weapon systems and military organizations, and they are exacerbated when either becomes more complex.

The message of this book is not that simple is always better. It is rather, that, under some conditions, complex can be worse—that complexity can actually increase the uncertainty that constrained organizations, especially military organizations, face. This discussion has particular importance for Western societies that are increasingly using complex technology as a substitute for large standing forces. Both the draw and the drawback of modern weaponry is its complexity. The more complex weapons have great appeal because under the best of conditions their capabilities offer enormous possibilities for survival and success in battle. Because these capabilities are so interrelated, however, there is much we cannot know about how the machines and their users perform under the more near to normal "worst of conditions" found on the battlefield.

The insights in this book have an importance far beyond the confines of any particular military service. Several audiences need to understand how it is that complex machines induce complexity in already constrained organizations. Students of national security will find here some explanation of the friction in modern operations and the perverse outcomes of war. Defense analysts will find a method for evaluating those assumptions about technological advantages before they become part of the organizational value system and before the outcomes are hot potatoes in the political debate. Participants in the debate over military reform will find common ground that will provide a basis for understanding how even with the best of intentions an expensive program can produce a marginal weapon. And organization theorists will find useful contributions to the process of evolving theory in a world of organizations that are vigorously seeking complex technologies. The story herein also has implications for the nuclear community, for which perverse outcomes either in weapons or in organizations have truly global consequences.

[ix]

The story this book explains is also crucial to understanding the systems around us. It helps make possible the identification of conditions under which an increase in complexity is likely to be a problem. The mid-level theory and tests herein will enable military scholars, organization theorists, and policy makers to arrive at a common and consistent understanding of technological and organizational complexity and its implications for their decisions.

The primary intellectual influences on this book are those of Todd R. LaPorte and Gene I. Rochlin—incisive and unconventional thinkers whose discussions with me stimulated new thinking about the problems of organizations, technology, and public policies, especially defense-related ones. Both constant friends and colleagues, their encouragement and critical analysis were essential to this unconventional work. The quality of the book is due in great part to their unswerving support; the shortcomings are mine alone.

My thanks must also go to the Logistics Management Institute (LMI) of Washington, D.C., for enabling me to extend the knowledge gained in its projects to make a broader theoretical contribution. I also express my gratitude to the Institute on Global Conflict and Cooperation of the University of California at San Diego, whose fellowship support was essential to the completion of this work. Special appreciation is also due to the Institute of Governmental Studies at the University of California at Berkeley, which provided me throughout this effort with an intellectual home, a friendly library, and access to a copier—three critical ingredients for progress. Several other people also deserve special mention. Edward D. Simms unstintingly provided insightful guidance in and around the culture of the defense community. Lucas Bragg gave me an invaluable education about practical aspects and difficulties of weapons engineering. And the constant encouragement of Gene Narragon was indispensable in my efforts to keep my sights on completing the book. Equally important was the personal friendship that each of these people offered and I gladly accepted.

Other friends and colleagues in Washington influenced my understanding of weapon systems development. Frank C. Spinney introduced me to the subtleties of procurement and budget decisions.

Frank McDonald and Pierre Sprey, two people whose experiences in understanding organizations spanned more than two decades, often stepped in to help me out of mazes of my own making—Frank with his common sense and methodical approach to the study of maintenance and its problems, and Pierre with his patience and years of knowledge. Pierre once explained his twenty-year struggle for less expensive but more effective weapon systems by saying simply, "It would be dishonorable to quit." In military issues, lives are at stake and honor requires such dedication.

The contribution of my spouse, Dennis A. Lowrey, is beyond my pale words to describe. Army officer, scholar in his own right, unswervingly faithful friend, dedicated and critical editor, Dennis was there to rally the flag and find the computer program, book, or missing premise when my resources were spent. Patient during long hours poring over some convoluted phrasing, he managed not to lose that patience when my eyes had crossed and another figure had to be hand-drawn at the last minute. This paltry set of words is an understated paean to his love, collegiality, and brilliance.

CHRIS C. DEMCHAK

West Point, New York

Military Organizations,
Complex Machines

". . . chance favors only the mind that is prepared."
—Louis Pasteur

[1]

The Role of Complexity

In the late 1970s the message from senior members of the United States Army to its manager was clear: battlefield uncertainty must be met with "maximum use of technical advances." In 1981 the army chief of staff stated in a speech: "We are not likely to be able to buy hardware in quantities equal to those of the Soviets. Consequently we need advantages in the equipment itself and in the means and methods of its employment." In 1982 a senior U.S. Army leader said, "We are moving to make the force more lethal and less personnel-intensive through the exploitation of high technology."[1]

About the same time, a paradox emerged. Program managers were developing the army's new supertank, the M1 ABRAMS, which was intended to be simple to operate and to repair. However, it could *not* be repaired without a significant amount of help from the contractors, and some symptoms mystified even the contractors' engineers. The Army responded by contracting for *another* machine—"alternate test equipment"—which was supposed to tell the repairer what the problems might be in the bigger and unexpectedly complex supertank. But as time went on, the smaller machine could not be repaired without special help, and the Army created a new job specialty for that task as well. Those who operated the machine were trained separately; those who maintained it were given new job categories. Finally,

---

[1]See, respectively, Guthrie 1979:53, Meyer 1984:198, and Richardson 1982:252.

faced with enormous unexpected costs, the organization began to monitor the parts for the tank on a worldwide basis. In sum, the tank that was meant to be simple, cheap, and reliable turned out to be complex, expensive, and surprising.

This paradox of complex technology has been around as long as complex machines themselves, but a widespread tendency to view new technology always as beneficial, or at least neutral, has obscured understanding. A traditional focus on soldiers and organization—not on equipment—exacerbated the situation by leaving the army organization unprepared for the almost inevitable consequences of introducing complex equipment on a massive scale.

Although scholars have long recognized the significance of the ways in which plans, places, people, and machines interact on the battlefield, people have dominated in the retelling and analysis. For much of history, the technological differences among warring forces were minor, and opposing forces generally displayed parallel technical developments. Occasionally a technological advantage was available for years before it was successfully implemented in the military forces. Even the use of innovative tactics or careful choice of ground—both human actions—was viewed as more decisive, and hence the classical scholars did not tend to view technology as a significant contributor to the problems of operations.[2]

This presumption continues today, even among the most exceptional scholarly thinkers. Concerned about the chaos of war on command and control systems, one prominent Israeli scholar suggested that there might not be any chaos in a war of machines against machines; the implication was that only humans cause or promote the undesirable or unexpected. Others attribute the problem to the human tendency to purchase the wrong thing through either political or sociological pressures.[3] In each case, however, human flaws are blamed.

---

[2]Such greats among military scholars as Clausewitz and Sun Tsu do not discuss technology. Jomini is unique in that he discusses logistics in his prescriptions for war. See Handel 1986, O'Connell 1989, and Van Creveld 1989 on the reasons for this historical neglect, and Ellul 1964, Winner 1977, and Steele 1983 on the tendency to view technology as neutral.

[3]See, respectively, Lanir et al. 1988:100, Kaldor 1981, Fox 1974, and Fallows 1981.

[2]

Modern warfare has brought the role of machines to unprecedented levels. Today's military organization comes to fight altered significantly by the machines it chooses to use. Fighting capability is altered by the new equipment, but the structure of the organization itself adapts to the innovations too. Plans, people, and places are all affected, and sometimes driven, by presumptions about and the requirements of the machines.[4] This interaction of organization, environment, and tools can cause maddening outcomes, for which there are, to date, still few explanations.

Although an increase in the complexity of military organizations is clearly concurrent with an increase in the complexity of military weapons, the connection between the two increases has rarely been explored. In the literature of military scholarship, advances in technology are seen as almost universally beneficial, at least to the warring side that uses them first. That new weapons and better equipment have a phenomenal effect in battle is clear but that they might also be responsible for some of the problems is widely ignored.[5] This book explains how an organization's efforts to cope with the unknowns in complex technology, which are successful in peacetime, can in wartime produce slower and more fragile tactical forces and perhaps disastrous consequences.

SURPRISE

The origins of what seems to be a paradox of complex technology lie not in the machines or in the humans' organization per se, but in the complexity of both. The most important characteristic of complexity is that it produces *surprise*. Wherever a complex system is found, and in whatever form, its outcomes are usually not quite the expected. At some point the various relationships in

[4]Some machine shortcomings noted as malfunctions in peacetime maintenance logs, such as broken headlights, would not make the equipment incapable of fighting in wartime. These are not the kinds of failures under consideration here. But major malfunctions of combat capabilities, especially in the electronics, *are* shortcomings that increase the perceptions of uncertainty and drive organizational responses.

[5]Exceptions are increasing. See Kaldor 1981, Van Creveld 1989, and O'Connell 1989.

a complex system undermine even the most careful programming of every component. Minor variations sum unpredictably into an unforeseen outcome (Kahn 1968). Highly touted expensive machines can simply stop firing, garble transmissions, or aim at the wrong target. While unpredictable results can be either good or bad, negative results are noticed more often and can appear even malicious. Those negative results are the ones that the members of the organization remember, and in their responses the complexity of machines affects the organization as a whole.

Surprises are particularly troubling in large organizations, such as armies in battle. A military is a large, societally important organization for which the consequences of "not knowing" can be extremely costly. Like others, militaries face a demanding environment, but poor military performance directly affects the survival of the organization's members: a poor army dies. Furthermore, its operational setting—the battlefield—includes an enemy eager to deliberately cause disabling surprise by any means, so military managers have an unusually strong interest in controlling behavior in highly dynamic and stressful situations. They are eager to acquire machines that are likely to make performance more predictable.

Unfortunately, without special efforts to mitigate the effects of the uncertainty of the machines, the level of surprise in military or similarly constrained organizations can increase. Complex machines come to the organization with their own tendencies toward unpredictability and are introduced into hierarchies that are usually large-scale and that face a certain irreducible level of surprise inherent in their organizational complexity even before the new machines are added. In their zeal to reduce uncertainty in the organization, managers have consistently interwoven the two types of complexity before all the consequences are known.

The results are often quite surprising, and a perverse cycle develops in the organization. As the new equipment widens the organization's relationships, individual variations in behavior aggregate unpredictably. Because the initial goal was greater control, new surprises are likely to force even more pronounced managerial efforts to restrict new offending behaviors. Managers may

[4]

even acquire more machinery to help control the behaviors. For example, they may add monitoring, guiding, and restricting relationships that increase the complexity of the organization as each new link is added. The result can be a rigidity in which flexibility is lost[6]: Members of the organization respond slowly or inaccurately to unexpected conditions, members at key nodes make mistakes, choosing the wrong response, based either on tradition or on too much or too little information. As these minor mistakes accumulate, the possibility that there will be unpleasant surprises increases.

Surprise is particularly harmful to battlefield performance. For military scholars, the costs of surprise are well-known. The eleventh-century Chinese military scholar allegedly responsible for much of Mao's military genius, Sun Tsu, urged military leaders to promote surprise, and hence disorder, in enemy forces. Surprise and disorder would lead to the enemy's defeat, but a threat for the enemy was equally dangerous to the friendly forces. According to Sun Tsu, 90 percent of a war was won or lost before the actual battles began. The successful leader prepared extensively in advance in order to avoid being surprised (Wing 1988).

To explain the hardships of battle operations, the Prussian military officer Carl von Clausewitz suggested a concept from physics: "friction." In the early nineteenth century the set-piece battles of the previous century (fought largely by mercenaries) ended. Wars became more dependent on factors that are less clearly under the control of the generals, such as the actions of individual soldiers in Napoleon's mass army. "Friction" was the explanation for the unexpected outcomes of the new warfare: "Countless minor incidents . . . combine to lower the general level of performance [in battle]."[7] Commanders often faced the unexpected—things operating differently from what was planned— and had too little information or time to come up with new plans.

---

[6]A society dominated by such organizations would resemble the Chinese bureaucracy before the 1949 revolution, with the resulting stagnation (Schurmann 1968:220).

[7]See Clausewitz (1976:119, 102). Clausewitz also contributed to the current difficulty with complexity by stating that friction "is a force that theory can never quite define" (1976:120).

[5]

For Clausewitz, surprise and friction were interrelated and common on the battlefield.[8]

"Surprise therefore becomes the means to gain superiority" (Clausewitz 1976:198). In the modern army, the organizational structure produces both friction and surprise, and a military organization can impose surprises on itself, giving the enemy a certain advantage. The need to coordinate the many multiplying relationships produces disorder in the face of crises. One military scholar claims that the multiplicity of headquarters and the resulting extraordinary need for information were among the chief causes of the difficulties the United States had in Vietnam. The amount of time required to coordinate operations lowered the speed of U.S. responses and gave the enemy opportunities for disruption that Sun Tsu would have avoided (Van Creveld 1985:258).

If an organization's structure can itself cause hardships and defeat, and the new machines can change the structure, then those new machines can alter the possibilities for friction and surprise. Hence, all things being equal, the organizational conditions for high or low levels of friction and surprise are directly related to the machines. Complex machines are themselves more likely to cause surprise as well as to induce certain managerial responses to that surprise. The complexity of the machine influences the organizational conditions for greater or lesser hardships in wartime.

CHOOSING OUR CASE STUDY

There are strong theoretical and practical grounds for selecting a military organization as a case study to use to explore the general phenomenon of the paradox of complex technology, and specifically the effects of complexity. The military is an organiza-

[8]Clausewitz's concept of friction is evocative, but it is not elaborated sufficiently to be a useful guide for organizations. Anything can cause friction; anything can come from friction. Conceptually wartime friction becomes a kind of cosmic bad luck. In this view, friction on the battlefield produces surprises, and little can be done to reduce either the friction or the surprises. Yet some armies have had fewer problems with friction than others, while opposing each other in the same general environment. Hence, the components that produce friction and surprise must be understood.

tion that is seriously constrained. Generally short of funds, time, and operational choices, it is very sensitive to surprises. Certain information is needed to adapt appropriately to the unexpected, the military is not able to buy, wait for, or reorganize or operate successfully without it. The organization cannot simply sell off an element rattled by complexity, nor can it buy enough to carry a spare for whatever does not work. Physical limits bar rapid, frequent, technical rearrangements, but lengthy downtime is also not possible. Finally, working conditions do not allow the constant trial-and-error process necessary to limit the number of surprises (Landau 1973). In sum, if complexity increases and critical information is scarce, the organization will show the effects clearly.

Several organizations fit this description,[9] but a military organization is a particularly useful case study because of its self-conscious error avoidance. It exhibits a strong need to have the right information at the right moment in the right location in wartime. In addition, there is a strong organizational imperative to predict outcomes by collecting information that is critical to success and by trying to avoid mistakes. Managers seek to gain control on the basis of information and in response to anticipation of possible mistakes are apt to change structures uniformly. If harmful surprises occur in spite of these efforts to avoid or mitigate them, then the effects of complexity on a system are demonstrated.

A military's size and nature offer several research advantages. First, as a large-scale organization, the military makes its organizational changes explicit in order to communicate with all the elements concerned. The military organization is most likely to document allocations, to emphasize formal internal structures, and to standardize rules. Uniformity in responses is not possible when orders over the length and breadth of such a system are verbal. These organizations plan extensively in advance and formally propose, assess, accept, and implement changes in organi-

---

[9]This group of organizational limiting cases includes air traffic controllers, electric power companies, nuclear power generating systems, large banks, air forces, navies, and armies.

zational operations. When increases in complexity occur, responses are most recognizable in a military organization.

Second, these organizations purchase modern machines to perform critical functions in an effort to reduce wartime surprises. Hence, the acquisition of machines is tied to organizational certainty. To avoid mistakes, the organization is altered to get the benefits of the new equipment. Given the comprehensive planning and standardization, the organization is affected considerably if the accepted understanding of technical complexity does not meet the actual uncertainties of operation. A new weapon system is a convenient, critical, and identifiable source of increased technical complexity whose surprises are likely to induce some kind of visible organizational response.

Finally, militaries make good case studies because of their responsibilities to society and their demands on national resources.[10] Explaining the effects of complexity on such organizations has value beyond the academic literature. The failure of a large corporation is disruptive, but the markets generally adjust. When a military service fails to perform as expected in wartime, however, the nation as a whole is at the very least psychologically distressed. Argentina's poor military performance in the Falklands precipitated political upheavals, as did Germany's defeat in both world wars, China's occupation in World War II, and Russia's failures in World War I. In the nuclear age, conventional military failures can provoke use of nuclear weapons and produce disastrous consequences.

As contemporary military organizations, along with their fellow

---

[10]It is useful to make a methodological note about case studies in general and the structure of this one in particular. This research is what Yin calls in his excellent work on the subject both a critical and revelatory case study involving embedded units of analysis. In other words, not only does this case test a theoretical approach to complexity, it also reveals material that might not otherwise be accessible or understood. Embedded in the single case are independent, control, and dependent variables—respectively, the more complex M1 tank, the less complex M60 tank, and the complexity of the organization (Yin 1984:47–50). The research strategy achieves substantive validity through a basic theoretical approach moving from hypotheses to evidence, conceptual validity through a combination of parsimony and scope, and methodological validity by emphasizing realism first and generalizability second. For an excellent discussion of the strengths and shortcomings of these research strategy decisions, see Brinberg and McGrath 1985:44–70.

[8]

organizations, rush to acquire computers, the effects remain unclear. The organizational results of increased technical complexity present pressing questions for students of military problems as well as for organization theorists. Research on highly technical organizations, such as the U.S. Navy's Polaris program, nuclear command systems, and current research on technical organizations that require high reliability has produced some suggestive results.[11] It appears that dramatic changes in organizational interaction and structure are needed to make operations with highly complex machines successful.

The goal, then, is to have authors, actors, and observers alike better understand the nature of complexity and its interactions among machines, organizations, and their environments. This book uses the case of the U.S. Army's M1 ABRAMS main battle tank as an example of events that produce the organizational surprises and the likely consequences in wartime. The U.S. Army is an excellent choice for this study because it is a large-scale military organization with a strong interest in internal certainty that has over the last decade been aggressively pursuing a massive program to equip its personnel with advanced technical machines. In 1981 the U.S. Army announced plans to acquire four hundred new systems over the next five years (Meyer 1984; AUSA 1982:23).

In addition, the Army has special research value as a relatively nontechnical organization; responses to complex machines are highlighted more than in other more "machine-oriented" organizations. The U.S. Army historically focused on the soldier, not on the equipment. The Army operations field manual FM 100-5 (1986:5) states: "Wars are fought and won by men, not by machines." The Army is not in the habit of dealing with the precision requirements of machines, and Army personnel tend to assume that surprises are the result of failures by the *people* in the organization, even if the source of a surprise is the complex machines the people are using.

For example, maintenance problems are often blamed on poor leadership: "Commanders who complain about how little their

---

[11]See, respectively, Sapolsky 1972, Bracken 1983, and LaPorte et al. 1986.

junior mechanics or supply clerks know... fail to realize that the great majority of soldiers' skills are developed in the unit, on the job, by qualified supervisors" (Meyer 1984:202). In other cases, the problems may be blamed on paperwork: "In general, I am not happy with where we are today [regarding maintenance]. But leaders can only go so far in compensating for the laborious, paper-intensive system that is in place today" (Meyer 1982:24).

When faced with machine-induced surprises, managers tend to respond to difficulties by rearranging the *people* around the machines, thereby in visible ways altering relationships to accommodate complexity. This tendency is further reinforced by a pervasive approach called the "can do" attitude. It is considered highly inappropriate for officers or senior enlisted personnel to complain about a situation if the listener can do nothing to change it. Hence, in the organization stoic resolution—often called a "can do" attitude— is strongly reinforced. Senior officers who view the support problems of machines as inevitable and immutable display a "can do" attitude by finding profitable solutions in things that are clearly within the organization's purview to change, such as leadership and training.

As a military service, the U.S. Army is also likely to suffer self-consciously from an increase in complexity. Unable to knowingly tolerate substantial organizational error and risk the loss of lives, battles, or wars, the managers tend to reach out to avert what they perceive to be possible surprises. The decisions these managers make reflect what they think increases certainty. What is and is not changed indicates what is considered critical to performance. The sum of their decisions reveals the organization's understanding of complexity and its effects.

Finally, the U.S. Army is a significant institution in the nation's political system, and the political repercussions of poor technological choices and organizational adaptations can be substantial. They will be felt not only on the battlefield but also in the halls of Congress, where funding outcomes can be no less decisive.

When the U.S. Army began its long-awaited modernization in the mid-1970s, its managers tried to maximize the combat capability of a single machine. They sought a design that promised both more capability and more simplicity—and they developed the M1 ABRAMS main battle tank. The compactness of these capabilities also helped the budgetary crunch associated with massive modernization. With more capability per unit, the Army could safely buy fewer of the expensive machines. And the managers expected that the tank's operational simplicity would mean less costly and earlier front-line maintenance. More savings would be possible because of a reduced training level and fewer maintainers needed at the front. These assumptions were implemented in new maintenance doctrine and requirements along with the development of the tank.

The tank gave the combat forces new abilities and promised the tactical maintainer a less demanding job. Removing subsystems ("black boxes") was physically easier than in previous tanks. The goal was achieved, at least nominally. However, along with the new capabilities and apparent simplicity came great internal complexity, and this combination of internal intricacy and the military's need for certainty in availability brought about a cycle of positive feedback. With the machines, interdependence increased. Tighter strands of relationships reduced the ability to recover quickly from disturbances. As surprises rippled more easily along the growing web of necessary interactions, the tactical organization was faced with more unexpected outcomes, and a single disruption could affect more of the organization. The complexity of the weapon system drove increased complexity in the rest of the organization, producing a more complex and less robust tactical military organization.

Technological and Organizational Complexity

The argument of this book has two main parts. First, complexity in critical machines induces increased organizational complexity in constrained organizations, a process that is a result of the various kinds of unknown and unpredictable outcomes inherent in complex systems. Scarcity of knowledge, inherent in complex systems, induces a positive feedback cycle in which managers' attempts to reduce the uncertainties of the new machines increase organizational uncertainty. Second, a more complex military organization will have proportionately greater problems in trying to operate effectively in wartime. The more complex organizational relationships distort knowledge identification and transfer mechanisms. Outcomes are therefore not only unpredictable but also likely to be highly undesirable.

Most of this book is dedicated to sustaining the first half of the argument because it is most likely to be contended. This chapter introduces the argument and provides the context for the rest of the work. The importance of complexity in certain systems is discussed as are the reasons for choosing the U.S. Army and its main battle tank as a case study. Chapters 2 through 6 demonstrate the relationship between equipment complexity and organizational complexity.

Chapter 2 presents a theory of complexity that explains the source of the problems organizations have with complex equipment. A lack of knowledge that is associated with all complex systems produces the unpredictable and undesirable failures in performance, called *rogue outcomes*. This shortfall is not only problematical but also to some extent inevitable and ubiquitous. The ability of each organization to adapt itself to this missing information in operations determines the importance of complexity for organizational performance. In particular, three circumstances encourage machine complexity to ripple through the organization: (1) the knowledge encapsulated in and needed to run the new machine is scarce in the wider society, (2) the machine is critical to the organization, and (3) managers have a sense of urgency about the time and resources needed for effective performance. The chapter also provides tests for increasing complexity that are used in the remainder of the book.

Chapters 3 and 4 present the key elements of the case study. Chapter 3 focuses on the M1 ABRAMS main battle tank and its predecessor, the M60A3 PATTON tank; Chapter 4 looks at the U.S. Army and its managers. The massive modernization program, undertaken in the late 1970s and 1980s makes the Army ideal for exploring the organizational effects of complexity. The Army's difficulties with complex modern equipment exemplify the current tribulations of other organizations. As a significant public organization, the U.S. Army is under much scrutiny and receives much hostility when its expensive weapons perform poorly. Other organizations that need to understand how equipment decisions can have undesirable consequences will also find this study useful.

Chapters 5 and 6 present the evidence for increases in organizational complexity in response to the new and more complex tank. Chapter 5 identifies the key Army managerial decisions that were based on commonly held assumptions about modern complex equipment, decisions that helped mold the conditions under which the tactical units would adapt to the tank's complexity. A widespread misunderstanding of complexity by the managers produced conditions in which increased equipment complexity resulted in increased organizational complexity. Chapter 6 details what happened when the tank reached the tactical units and the units began to adapt to the equipment. The shortfalls in knowledge prompted an ever-deepening network of special relationships and arrangements that became essential for the successful use of the tank.

Chapter 7 addresses the second half of the argument: the implications of such informal and structurally intricate support webs for the wartime performance of these units. Military history emphasizes the power of surprise to destroy armies, and the need for both speed and accuracy to survive these surprises. The military battle doctrine is the military professional's best guess as to how organizations should respond to the unknowns of wartime in order to be successful. Comparing the likely speed and accuracy of these new complex relationships to similar requirements of both wartime and doctrine is the best available test of the implications outside of actual combat. Particularly useful is the Army's new AirLand Battle doctrine, designed with the new modernized army in mind. Ironically, the new equipment envisioned by the authors of this doctrine produced a tactical organization likely to have severe difficulty meeting key requirements of the very doctrine for which the equipment was purchased. The result is a military organization dependent on scarce knowledge for success in wartime. Its battlefield performance is less predictable overall, and it will take an exceptional array of resources and time to assure certainty in the outcome of a major conflict with a determined opponent. The price of failure will of course be paid in lives lost.

Finally, Chapter 8 concludes the book and recommends possible solutions. Some things can be done to make complex equipment less problematical—for example, designs tailored to the wartime knowledge conditions of the organization, and more rigorous testing, would increase the information available about the machine before it is introduced into the tactical organization. In peacetime those factors could dampen the tendency for machine uncertainty to ripple through the managers' efforts at control. Also, a military-staffed production organization that emphasizes testable and reliable systems could be reinstated. Such an institution is likely to be slower in innovation, but would also produce systems that tend to be less complex.

But there will always be limits to how much we can know about the operation of a complex system,[12] and complexity will always cost a good deal. The closer the organization attempts to get to the limits of knowable outcomes, the more the necessary knowledge will cost in time and money. Hence, constrained organizations will always have difficulty buying the knowledge needed to operate a complex machine successfully and reliably.

Militaries, in particular, are constrained organizations for which complex weapon systems present particular difficulties. If such systems are introduced into critical functions, the trade-off can be posited as a choice between extraordinarily large military budgets and losses in wartime. But the costs will be paid both ways. The new equipment will increase in cost as long as it is complex and scarce in the wider society. And aside from the sums paid in peacetime, unpredictable and undesirable outcomes endemic to complex systems will still exact a price in lives in wartime.

Societally significant organizations can least afford to be ignorant about factors that increase their own complexity and exacerbate their normal set of difficulties. Under some conditions, new complex equipment could be beneficial, but under other conditions there could be broad and reverberating harmful effects. Some mistakes have serious political consequences both for the organizations and for society—the former could be dissolved, the latter seriously disrupted. This book hopes to offer an understanding of complexity that will help avert these costs.

---

[12]In the 1991 Gulf War, U.S. military organizations had extraordinary redundancy in assets but still required six months of trial-and-error learning in order to operate successfully. See Chapter 7 for comments.

[2]

The Power of the Phenomenon

Complexity in dynamic systems is poorly understood. In many cases, reliable small components connected to a multitude of other reliable small components tend to produce much less reliable whole machines. We can multiply probabilities and derive an estimate of reliability, but we have difficulty explaining the processes of complexity. Furthermore, it is difficult to say when complexity is harmful and when it is not. We can observe practical truths but have no generalizable way of comparing them.

Academic and nonacademic traditions generally provide a poor understanding of complexity[1] and of its problematic output, *surprise*.[2] The complexity of a system is rarely explicitly addressed as an independent variable in a chain of events; even more rare is an actual operational definition.[3] Group-oriented disciplines, such as sociology and political science, tend to treat complexity as a

[1]For notable exceptions to these observations and good discussions of each, see Gleick (mathematics and engineering), Warsh (economics), Tetlock (psychology), Perrow (sociology), LaPorte (political science), and Erlich et al. (biology).

[2]I am grateful to Todd LaPorte for suggesting the term "surprise." A number of commonly used terms—uncertainty (Thompson 1967), chaos (Lanier et al. 1988), exception (Galbraith 1977), etc.—do not capture enough of my intended meaning. It is not a generalized lack of knowledge as in "uncertainty," or a decomposition of structure as in "chaos," or the unusual but possible expected event which is an exception. Rather, surprise is the human feeling of shock when one or a handful of quite unexpected outcomes simply appears. The outcomes are then surprises.

[3]In his discussion of complexity in theory, Wilson argues that although complexity is often reflected in the political theory literature under a variety of names, such as "diversity," the theory is nonetheless lacking (Wilson 1975:331). See also Galbraith 1977 as well as Kraemer and Perry 1989.

deus ex machina not amenable to manipulation or serious study. Such item-oriented disciplines as mathematics, engineering, economics, and psychology tend to view complexity as a pathology that develops inexplicably as individual elements aggregate and is to be avoided—and studied only so we can learn to avoid it. Biology embraces complexity as a stabilizing attribute: complexity is a good thing, and a system cannot have too much of a good thing.

However varied these approaches, they typically devote their efforts to subsuming, accounting for, or otherwise controlling for the effects of complexity, without directly explaining the phenomenon itself. For example, one noted scholar calls complexity a convergence of political struggles, a definition that may identify its troubling aspects but does not say what variables create the situation (Steinbruner 1974:15–16). Others prescribe correctives without a clear or cogent explanation. For example, much of the popular business management literature is anecdotal and guiding principles are offered in the form of lists and more lists. A correlate in the defense community is the "quality versus quantity" dispute that occupies much of the defense reform debate. Both sides declare complexity [in machines] to be neutral or bad and quickly move on to make prescriptions.[4] In sum, what is missing is an explanation of complexity that is theoretically consistent and operationally appropriate.

The widespread confusion that results suggests a fundamental difficulty in coping with the phenomenon of complexity and its symptoms, especially the surprises. The literature on complexity is incomplete; work so far has focused on some aspects of it but neglected the power of the phenomenon as a whole.[5] Complex

[4]See the final chapter of this book.

[5]Examples of conceptual difficulties with complexity and surprise in their various guises exist even in some of the better works. One well-known author observes that size and wealth decrease internal and external uncertainty for militaries without much explanation of why this intuitively problematical outcome might be true (Posen 1984:49). A second author suggests that a war without chaos can exist if it is solely between machines—which means no surprises or uncertainty (Lanier et al. 1988:100). Even Perrow's otherwise excellent book seems to muddle the concept of complexity by linking it to loose coupling and indirectly to both greater and lesser redundancy (Perrow 1984:280). The operational difficulties are obvious.

systems produce the unexpected with annoying regularity, but the fragmented set of approaches to complexity has often produced inconclusive research and little broadly applicable guidance. For example, redundancy is either a cure for or a cause of complexity, depending on the author (Landau 1969 or Lanier et al. 1988). The situation is exacerbated by the tendency of complexity to vary in its significance across systems.

Nonetheless, complexity's ubiquity and varying effects suggest that it is a distinct phenomenon open to research and to the development of a general model. With such a tool, we could examine common elements whose results vary according to conditions in the different systems. Such a model also would allow for rough comparisons between systems, even those as different as organizations and machines.

MULTIPLYING RELATIONS

Such a general model must first explain the most important characteristic of complexity: that complexity produces surprise. Wherever a complex system is found, and in whatever form, its outcomes are usually not quite what was expected. At some point, the mass of varying relationships in a complex system undermines even the most careful programming of every component. Minor variations add up unpredictably to an unforeseen outcome that can be either more or less desirable (Kahn 1968, Hofstadter 1989).

As an analytical term, "complexity," represents the structural intricacy of relations within a system.[6] When the components of these relations change, so does the system's level of complexity. Increases in these structural dimensions matter because they signal the increased difficulty of knowing what will come of the

[6]The term "structural intricacy" has been used to suggest the various components of this "complexity," but it is a synonym useful only for heuristic purposes. The word "structural" suggests visible hierarchy, but my intent here is to capture both formal and informal connections in a complex system. For this reason, I am using a hybrid of state and process descriptions of the system, commuting between literatures to achieve a workable synthesis. See Simon in Leavitt and Pondy (1973:644–73) for a more complete discussion of the distinction.

[17]

multiplying relations. Up to a certain point, the importance of complexity is relative to the critical knowledge available about the system. The more of these internal relations there are overall, the more difficult it is to acquire enough knowledge to know exactly how to respond to disturbances. Hence, increases in complexity are inextricably linked to increases in surprise.

The Knowledge Burden

Knowledge is essential for a system; without it, component parts would not know how to interact. For most systems associated with humans, operating successfully means ensuring that outcomes are relatively predictable. The *knowledge burden*, therefore, is the amount of information needed to ensure that outcomes of the system are not surprises (Galbraith 1977, LaPorte 1975). The more structurally intricate a system, the more we need to know about the intricacies in order for it to function according to the desired pattern.

In this book, "knowledge" is more than just information carried by an individual or contained in human learning; it also applies to machines. A spare part embodies knowledge. Complex machines require more precision in their inputs, and creating those inputs requires knowledge. Hence, the inputs too represent knowledge. A highly integrated machine will have less tolerance for error in the connecting cables between internal elements, so more sophisticated machines are needed, to produce the cables themselves, and therefore more knowledge is required for the machine to operate as anticipated.

Complex systems have large knowledge requirements both initially and over time. When a system first begins to operate, it faces a universe of possible outcomes, many of these outcomes will predictably occur, and many will prove to be irrelevant. These outcomes constitute the "knowns" about the system—over time the largest category of outcomes. The set of "knowns" grows as the system runs, creating a learning curve that varies from system to system. For complex systems it generally takes more time to accumulate enough knowledge to move significantly upward on the learning curve.

[18]

At any given point, there will always be outcomes that may be unknown, many of which are in principle knowable if the necessary research to identify them is done. Often, however, they remain knowable unknowns because pursuing the information was too expensive, deemed unimportant, or not recognized as necessary. For the complex system, the number of knowable unknowns tends to be large. A good deal of investment is required to know what to investigate most profitably among the wide array of relations in the large system. An interesting example is found in the nuclear alert system. For years, system operators took care to make sure the system computers were told to send the correct messages to the various alert centers, but they did not think to make sure the message they intended was the message actually sent. It took several extraordinary false alarms to make this clearly knowable unknown worthy of investigation (Perrow 1984).

Knowing all the possible outcomes of a dynamic system is theoretically impossible, so there remains a final subset of unknowns that are simply unknowable. No matter how much research is performed, some subset of outcomes will remain mysteries because a system cannot evaluate itself (Godel discussed in Hofstadter 1980, Nagel and Newman 1958:94). Therefore, the outcomes that remain after the knowns and knowable unknowns are subtracted simply cannot be determined in advance. These outcomes are unpredictable either in form or in frequency, or both. Included among the unknowable outcomes is the combination of four or five unrelated and undesirable incidents at the same time. For example, failures in bank operations could occur because a solar eclipse magnetically affected energy transmissions in a power grid network, overwhelming the surge protectors of the main computers of the bank just as backup systems were down for maintenance.

Uncertainty is the lack of necessary knowledge at the right time and the right place, and for the complex system unknowable unknowns produce a great deal of uncertainty (Galbraith 1977:36). A good example is the case of Three Mile Island, where malfunctioning warning lights and organizational patterns of behavior led through a chain of unpredictable events to a reactor

[19]

core meltdown. (Perrow 1984). While extreme, these examples indicate the multiple and unpredictable nature of the events in complex systems. The greater the number of essential relations among a large number of components, the greater the likelihood that improbable events will ripple through the system and produce an unpredictable outcome.

Surprise is therefore greatest in complex systems because the unknown outcomes are proportionately higher than in more simple systems. All things being equal, knowledge becomes more difficult to acquire as a system becomes more complex, especially while the system is in motion. If a disturbance can travel along only one path, it is easier to predict the outcome of the disturbance. If in a large array of relationships a disturbance can travel along any of a number paths and branch at each component it encounters, then any given disturbance can take any path. If many types of disturbances are possible, prediction problems multiply exponentially and rapidly.

Rogue Outcomes

A complex system is more likely to experience *undesirable unknown outcomes* (see Figure 1). In any system, each node can tolerate a certain level of variation in inputs. If at each node the variation is at its maximum, then the level of variation accumulates over a series of nodes, so that at some point the input is either not accepted or disrupts the nodes. A good example is the children's game of "Telephone." At the end of the line of whisperers, when the secret phrase is spoken aloud and compared with the original, the final phrase is usually markedly and comically far from the original. A complex version of this game would have lines of people branching off from each person in the original line, making it possible for the phrase to travel in unforeseeable directions as well as to change unpredictably.

In order to pass many nodes without taking a wrong path or becoming seriously distorted, communications and input/output relations in the more complex system must be more precisely controlled (Pierce 1980). Deviations are not accepted as well as in a less extensive system. Replacements for inputs or nodes must

[20]

UNDESIRABLE UNKNOWNS

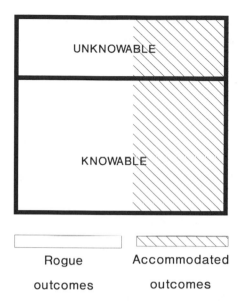

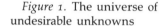

Rogue   Accommodated

outcomes   outcomes

Figure 1. The universe of undesirable unknowns

be carefully selected, so that they meet the precise requirements of the existing streams of inputs and outputs. This is particularly true of machine systems and increasingly true of human systems connected by machines. The complex system, which is likely to have numerous differentiated and interdependent relations, is less able to rearrange those tightly coupled relationships quickly or accurately when a disturbance occurs (Weick 1979, LaPorte et al. 1986).[7]

Complex systems are therefore prone to have small disturbances multiply quickly into large and nasty surprises. Undesirable unexpected outcomes form the "rogue set," a label intended to convey the sense of shock, often felt as betrayal, when a nasty surprise occurs.[8] The more unexpected and undesirable the out-

[7]The origin of the term "tightly coupled" is unclear, but in my reading of the literature Karl Weick appears to have first used it to mean strong interdependence between elements whose operations are critical to the organization. The term has a good deal of evocative power, expressing the consequences of high levels of interdependence (Weick 1979).

[8]I am grateful to Gene I. Rochlin of the University of California at Berkeley for suggesting this nicely evocative label in 1982. A rogue is someone of a mischievous disposition, dishonest and unprincipled. An undesirable outcome that comes as a surprise can easily be seen as a rogue (Oxford Univ. 1981:752).

[21]

come, the more unruly or roguish it appears. We call them *rogue outcomes*.

Rogue outcomes make complex systems difficult to operate reliably. In the computer industry, for example, the sources of these rogue outcomes are anthropomorphized and called "gremlins." They are so common that many computer systems are sold on the basis of the type and amount of maintenance support available when unknown outcomes occur. Major computer firms are known for their ability to provide technical support when something goes wrong with their machines (Kosakevic 1982). Despite a low probability of such rogue outcomes, they are feared because of the lack of knowledge at the time of the disturbance and the consequences of such a lack.

Accommodation

Rogue outcomes are generally unpopular. In human systems, the more significant the organization, the less tolerant the population is toward the unexpected, especially if it is deleterious. A great number of government agencies exist in order to make outcomes more predictable and undesirable outcomes less likely. For example, the Securities and Exchange Commission exists to monitor the stock market and keep the system from collapsing as it did in the 1920s. Other federal organizations exist to keep the rogue outcomes at a low level in the wider economy: the Federal Reserve Bank, the Federal Deposit Insurance Corporation, and the Federal Savings and Loan Agency.

However, as depicted in the shaded area of Figure 1, only greater knowledge lessens the effects of rogue outcomes, either by revealing the form or frequency of some previously unknown outcome or by guiding the systems' responses accurately.[9] When

---

[9]Whether the uncertainty is about the specific form or about the actual frequency of occurrence will have some effect on the nature of the accommodation required, but these are rarely two distinct kinds of response. For example, if the undesirable outcome is destruction of grain in a warehouse, if the rogue outcome is either marauding rats or deer who will come only at foreseeable intervals, then obstacles or deterrents that work at least once against each and can be replaced easily are necessary. If, however, it is certainly rats—but they could come at one or a number of times unexpectedly—then the obstacles or deterrents must be tailored

an undesirable outcome becomes less probable or less undesirable
due to changes in the structure of the system, it is called an
accommodated unknown. It continues to be unknown in form and/or
frequency, but the likelihood that it will disrupt the system
seriously has been dramatically reduced.

The goal of accommodation is to eliminate rogue outcomes or to
make them less roguish. For example, the Federal Transportation
Safety Agency investigates each major airline crash in order to
accumulate knowledge that will help make crashes more predictable and less likely. In principle, a subset exists for all outcomes
that are or have become predictable and are therefore known.
Unknown outcomes can be moved into this subset by efforts to
gain knowledge about the outcomes and by the structure's subsequent accommodation of the outcomes. This subset approximates
nil or very low uncertainty and hence is of little theoretical
interest.

Because accommodation of the rogue set in a complex system
requires a great deal of knowledge, the effort is both difficult and
expensive in time and money. For example, a significant proportion of public funds supports weather research in order to understand the probable outcomes of flying aircraft. The multiplicity of
factors producing the weather has been modeled, but running the
model itself is said to take as long as the actual weather events.

Accommodation in a complex system is expensive, and moving
the costs around does not necessarily reduce them. Knowledge is
scarce, and in a market economy scarce goods cost more. Perhaps
we could make every pilot a weather expert, but while that would
make the knowledge less scarce it does not necessarily reduce the
overall costs. The training itself is costly, and the new experts still
need data to calculate. A system of disseminating knowledge to
the experts must still operate, and the training must be updated
periodically as well. Therefore, the costs of accommodation in a
complex system are still high.

---

to rats and reinforced for multiple unforeseeable attacks. In either case, the
responses are generally similar and overlap in many particulars. The obstacles and
deterrents for predictable rat *and* deer marauding will be similar to those against
unpredictable rat attacks. Needless to say, it is most difficult to accommodate the
undesirable outcome when *both* form and frequency are unknown.

Even if all of the knowable unknowns are revealed, the rogue set would still consist of unknowable unknowns. Therefore, a system's designers and managers must decide how to live with the lack of knowledge. Owners of machines buy maintenance contracts from the sellers of machines, and some organizations (like banks and militaries) usually lean toward risk-averse responses and, if they can afford it, buy more training and machines in order to have backup systems in place when disruptions occur.

It is conceivable that in some circumstances rogue outcomes can be accommodated automatically by changing the nature of the knowledge needed.[10] For example, alternate energy proposals of the late 1970s provided systemwide accommodation against widespread power failure. Each component of the overall energy system—windmills, passive solar collectors, or co-generation plants—would not be highly reliable individually, but because of the number and dispersal of these components, the system itself would be robust. Rather than increase the precision of each node, the proposals assumed more highly redundant nodes but fewer pathways for disturbances to follow. Hence, such outcomes as failure of ten power generating sites are automatically accommodated and become irrelevant. No further resources need be spent gathering information on this outcome.

Such accommodation may be less expensive in routine outlays over the long run, but scarce knowledge will still be costly. In this example of the dispersed energy nodes, the research needed to find the best distribution of these nodes can be extremely costly—perhaps less than a failure, but nonetheless expensive.

The barrier of research expenses encourages larger rogue sets in complex systems. Testing material is expensive and often avoided in order to reduce costs during initial stages. Unknowns that

[10]Note that the set of accommodated outcomes encompasses some unknowable outcomes, which, though unknowable, can inadvertently be accommodated and often are. These are unforeseeable outcomes that are mitigated or do not occur because some unrelated feature of the structure happened to have a salubrious effect. For example, a teacher could discover that certain drills intended to teach the students x also left them more effective at performing something unrelated. For the rest of the discussion, however, these overlapping serendipitous occurrences will be largely ignored.

could have been known and perhaps automatically accommodated by early changes in design are often neglected until the system is on-line. A good example is NASA's decision to save money by testing the two key mirrors for the Hubble telescope only independently, not together. When the telescope was put in space, defects in the mirrors, which testing could have revealed, eliminated the value-added of the space telescope over ground systems (Booth 1990:32). The millions of dollars spent in producing the Hubble were effectively wasted. Automatic accommodation of the rogue set in a complex system is difficult to achieve; generally the outcome is left to occur before the required knowledge is sought.

Accommodation can also be graduated. Certain types of structures make the unknown outcome less likely to occur or, if it occurs, less undesirable and more neutral. In the case of the alternate energy system, the accommodation made the undesirable quality of the outcome diminish while not altering the likelihood that a node will fail. And the presence of spare parts, for example, does not reduce the likelihood of a broken tank, but they do reduce the likelihood of an *unrepairable* broken tank. Any set of problematical events that can be solved by inserting a spare part does not have to be evaluated for other solutions as long as the correct part is there. The needed knowledge—represented by the spare part—is present.

For accommodation to reduce the rogue set, the appropriate knowledge [for the system], must be available in the right form as well as at the right time and place. In the more complex system, this specificity makes accommodation more difficult. A common example is the calculator that operates from the wall socket *if* a specific transformer designated for that machine is available; the transformer is an unavoidable cost if we want to run the machine off the wall socket. Often the company's engineers have made the connecting plug or the wattage unique to that machine, so that no transformers from other machines will correctly transmit electricity. Hence, the machine-specific transformer represents critical knowledge because the machine cannot operate under the desired conditions without it.

Any scarce good has a higher price. In the wider economy,

[25]

information about transformers may not be scarce, but the form of that information necessary for this particular machine is. For the purposes of running that machine, the box of transformers we have accumulated from other machines is useless. The maker of the recorder or radio planned to make a profit from the transformer, knowing that once the consumer committed to the main machine, buying the additional item would be nearly unavoidable within reasonable cost.[11]

What is or is not scarce knowledge varies even more in a complex system, making accommodation more difficult to achieve. In the example of the calculator, a blackout will disrupt operations even with the transformer, so a supply of batteries is a partial accommodation of the rogue outcome. As long as the batteries are of the correct type and readily available to someone who knows how to insert them, a blackout is significantly accommodated in terms of its disruption potential. When a blackout occurs, then, batteries represent the critical knowledge, not the transformer. But in a complex system, knowing in advance what will become critical knowledge only when the crisis occurs is especially difficult, because the form and frequency of the rogue outcomes are unknown.

Accommodation of an unknown does not necessarily mean the unknown has become known. By definition, for example, unknowable unknowns cannot be known. Only by experience can students of general systems theory learn that a system with many potential outcomes also contains outcomes that cannot be known by inspection of the system (Nagel and Newman 1958). Hence, some accommodating of unknowns can in principle be done serendipitously without first acquiring knowledge. Experience has also shown that the greater the potential number of outcomes, the more difficult it becomes to achieve this serendipitous quality. More knowledge is required to achieve the same level of

[11]In the computer industry these extra items are called peripherals. First-time buyers of inexpensive machines are often misled by the price of the computer itself and unaware that an additional one-third or more of the purchase price must be spent to use the machine as they intended. Once the computer is purchased, the manufacturer is in essence in a monopoly position and can charge high prices, making a significant and disproportional amount of the overall profit on the sales of peripherals.

accommodation when moving from a system with few outcomes to one with many outcomes (Wohl 1980). By indirect connection, then, accommodation becomes more a function of knowledge the more outcomes in the system there are. In the remainder of this book, therefore, *accommodation* in complex systems will be assumed to be a synonym for having appropriate knowledge about the outcomes in question.

Complexity makes a system more surprising by increasing the number of potentially dysfunctional unknowns, or rogue outcomes. It is more difficult to design a structure to accommodate the undesirable unexpected outcomes because more knowledge is required to do that. In addition, more resources will be needed to acquire this knowledge, partly because it is scarce and partly because the theoretical limits of knowledge inhibit total accommodation of rogue outcomes. Hence, in the structurally intricate system more outcomes remain unknown and closed to accommodation than in simpler systems. The size of the rogue set is roughly inversely related to the knowledge needed to accommodate undesirable unknowns. Complexity makes a system's overall performance more "roguish"—more uncertain.

Measuring

Complexity is difficult but not impossible to measure. Obvious contributors to intricate relations are the individual elements involved. Indeed, the common understanding of complexity tends to focus on the sheer number of components involved in the system. Equally obvious, but less commonly acknowledged, is the contribution of individual variety to the intricacy of these relations. One hundred people in a system may or may not act differently from twenty people in the same system, but one hundred people performing ten different jobs in the same system surely act differently from one hundred people doing the same job. One key difference is that they may or may not need one another in the same way as the original twenty. Hence, not only diversity but also interconnectedness contribute to the level of complexity of these relations.

The complex system is therefore more than a two-dimensional

[27]

matrix of individual components and categories—it has a third dimension of direction and intensity of interdependence.[12] As these three variables change, so does the complexity of the

X system. In principle, then, complexity can be measured by an index made of three elements: the number of components (N), the degree of differentiation among the components (D), and the degree of interdependence among the components (I), especially the critical ones (LaPorte 1975:10).[13]

In operational terms, N is straightforward; it is the number of the components or nodes to be considered. In contrast, the measure for differentiation (D) must capture not only the number of different skills or functions but also their relative representation in the system. The common understanding of differentiation is "specialization," but differentiation is a broader term. If there are ten functions in the organization and only two people, then if both people performed five different functions each with no overlap between them it would be difficult to say that each person specialized. Yet the differentiation exists. The more the groups deviate from the average, as weighted by their relative representation, the more differentiated the system.

Finally, my proposed measure of interdependence (I) is the average of the minimal number of connections with other nodes that each individual node or component needs in order to operate successfully. The addition of any node, irrespective of its position in the system, constitutes some contribution to the overall inter-

---

[12]Winner (1977) warns that complexity should not be used as an independent variable because it is heavily interwoven with the size and rate of change in the system in question. However, although size is a component of complex relations, size alone does not determine the extent of the complexity (see Dewar and Hage 1978 for a discussion of the role of size). In addition, the rate of change in the numerical values of a system does not necessarily alter the structure (parameters) of the system. Averaging structural values over short segments of time is an acceptable way to allow for change. In the mathematical processes of integration-differentiation, the rate of change is measured by the distance between each identified point. If, as this distance becomes smaller with each additional point, the basic structure (types of parameters) changes, then these mathematical formulas could not be successful. In principle, they cannot be, because the number of points is infinite, but in practice the processes are used extensively and successfully. The operational definitions used in this work are meant only for similar analytical purposes.

[13]The number of components is labeled C by LaPorte (1975) and N in this work. The two are equivalent in usage.

[28]

dependence. Placing highly interdependent nodes at the bottom of the organization does not necessarily reduce the interdependence; each node is presumed to exist for a reason, and that reason is important to operations. Therefore, as a first approximation, the average need not be weighted. The figure does not accommodate some notion of superfluity. Further work beyond the scope of this book is needed to refine this or some other relatively simple calculation of interdependence.[14]

The relationship between the three proposed measures is not precisely delineated because the precise mathematical nature of the relationship remains daunting.[15] Similarly, if a single node is becoming internally more complex, how one extends that information to the implications for the entire system is not clear. A number of years ago a similar problem confronted economists in the relationship between capital and labor. Although the precise mathematical relationship between the two is still not identified, a substantial theoretical literature and voluminous research studies have developed. The same thing may happen for studies of complexity. For the purposes of this book, then, any change along one indicator without compensating movements in one or both of the other indicators will be taken to signify an alteration in the overall complexity of the system.[16]

It is important to note that there is no absolute threshold that

[14]A shortcoming of this proposed measure is that there are infrequently activated but critical connections. Ideally, each connection to the evaluated node would be weighted by a measure of how critical it is. Interdependence for the entire system would then be the average of the minimal number of weighted critical connections per node over the total number of the system's nodes. Producing a persuasive mathematical calculation of such criticality factors is difficult because identification of this minimal number of nodes would always be subject to the perspective of the individual assessing the crucial connections. Infrequent but critical connections are easily overlooked. It is often easier to describe the network of nodes than it is to establish criticality. Although primitive, the rough measure described here is conceptually similar to those used in engineering analyses (Bragg and Demchak 1982).

[15]For a tentative discussion of the kinds of mathematics involved in creating an integrated mathematical portrayal of complexity, see Demchak 1987. See also Casti 1989, *Administrative Science Quarterly*, and several popular computer journals for other approaches to the mathematical description of complexity.

[16]Landau (1969) suggests that redundancy is one way to increase the knowledge available in a given situation. See Demchak 1987 for a discussion of the relationship between redundancy, slack, and complexity.

separates the simple from the complex. The level of complexity in any system is indicated relatively by comparison either with other organizations or with itself in the past. Comparatively speaking, the more complex the system the larger the values for each indicator, that is, the greater the numbers, the differentiation, and the interdependence in the system. Correspondingly, these indicators can be used to demonstrate changes in the complexity of either machine systems or human systems, as long as the appropriate correlates to N, D, and I in the system are identified. If any system increases along all three of the components of complexity, it has become more complex.

Finally, using changes in an organization's structure as evidence of organizational response to complexity seems reasonable. Structure presents the bounds of action by establishing territory, allocating responsibilities, and at least nominally, channeling information. Some students of organizations have suggested that the real dynamics of an organization occur in its informal groupings. However, despite the form or variety of the informal groups, the formal structure of the organization inevitably plays a strong role in orienting both the participants and the researcher (Ranson et al. 1980). In any specific instance, informal interactions among certain personalities may dominate the outcome, but formal structure is more likely to dominate across a range of possible outcomes.

From a policy perspective, changing the informal groups throughout an organization is very difficult, while changing the formal structure is much easier. Organizations reorganize relatively frequently, and these changes have consequences for the informal groupings, their resources, their survival. etc. The significance of structure is particularly high for large-scale organizations in which the documented structure is used to direct resources and responsibilities as well as to orient new personnel. Formal structure indicates the hypotheses the organization holds about its technology, its environment, and the requirements for survival. Therefore, changes in formal structure indicate the direction and magnitude of organizational efforts to learn from these hypotheses.

[30]

As systems composed of human beings, expectations, and aversions, organizations have long been concerned with certainty, knowledge, and performance. An organization is a social system that develops an internal imperative to survive as soon as members develop an attachment to it (Selznick 1949). Survival depends on the organization's ability to acquire new information and to learn from it. Key to organizational learning, however, is the organization's ability to adapt its goals, attention rules, and the search rules used to consider alternatives. This ability to adapt is affected by the organization's ability to respond to changes with the right decision at the right time (Cyert and March 1963). The appropriate response to difficult events may be alterations in organizational activities. If the organization finds it difficult to change what it monitors, what it attempts to accomplish, and what ranges of possibilities it entertains, it is likely to find itself in a fight for its survival.

Controlling Uncertainty

For managers of organizations, a lack of knowledge is a troubling situation, and their responses exhibit a bias toward greater control of events.[17] Few managers will let a disturbance with seriously negative consequences simply exhaust itself. Rather, they tend to take action, either before the anticipated disturbance or after it has begun, to control outcomes by reducing the number of exceptions that can occur in the organization (Galbraith 1977). The more uncertain the organizational circumstances, the stronger and more self-conscious the managers' bias toward control.

Sources of external uncertainty are the most difficult for manag-

---

[17]Organizations are inherently a response to uncertainty. Oliver Williamson studied the conditions under which organizations—which he labeled "hierarchies" —may be a better alternative than a typical market solution to the problem of bounded rationality. Organizations curb opportunism and help computational and planning activities by aiding communications and enhancing the convergence of expectations (Williamson 1975:255). Those are the reasons for the organization's existence, as well as necessary activities for the organization to survive.

ers to control. In preparing to respond to disturbances from the environment, the organization may choose among four options— all of which are time-intensive or resource-intensive. The first, *buffering*, involves establishing organizational positions dedicated to watching the external sources of uncertainty.[18] An organization constrained in money but not in time could choose this option; the organization must be willing to wait for the knowledge it needs. Conversely, the second option—*insulating*—means neutral- izing the effects of the external source's actions on the organiza- tion. An organization constrained in time but not in resources could insulate by buying and distributing resources in order to meet all conceivable disturbances. The last two options are *absorbing* or *destroying* the sources of uncertainty, both obvious and large consumers of time and money (Thompson 1967).

Without slack in time or money, controlling external uncertainty becomes extremely difficult. The four options (see Figure 2) are not reasonable ones for the managers in fully constrained organi- zations. Hence, while they may initially attempt to control the most external sources of uncertainty, in practice managers eventu- ally focus inward,[19] seeking control in the internal mechanisms of the organization.

This inward focus is a natural outgrowth of both the historical development of industrial organizations and the influence of the scientific management theory in the field of business administra- tion during the first half of the twentieth century. For example, early industrial development in Britain was based on the use of excess labor. Managers then focused on reducing labor costs internally when the firm faced financial uncertainties, even if the money problems were externally induced (Barnett 1986). Frederick

[18]Thompson (1967:67) calls these individuals "boundary spanners."

[19]In general systems theory, stable systems are those that return to equilibrium (Boulding, in Shafritz and Ott 1987). For an organization, a certain level of uncertainty, if accepted over time, can be viewed as equilibrium. Changes in the environment constitute increases in external uncertainty, raising the level of overall uncertainty at least temporarily. If that level of uncertainty does not subsequently drop back down, managers will try to return to the original equilibri- um level by forcing drops in the internal level of uncertainty. This is in essence a balancing tendency that has been variously labeled over the years. One particular- ly well-known label is "cybernetic" (Steinbruner 1974), but there are others, such as "homeopathic." The notion is a common one.

Figure 2. Time/money constraints and organizational responses to environmental uncertainty

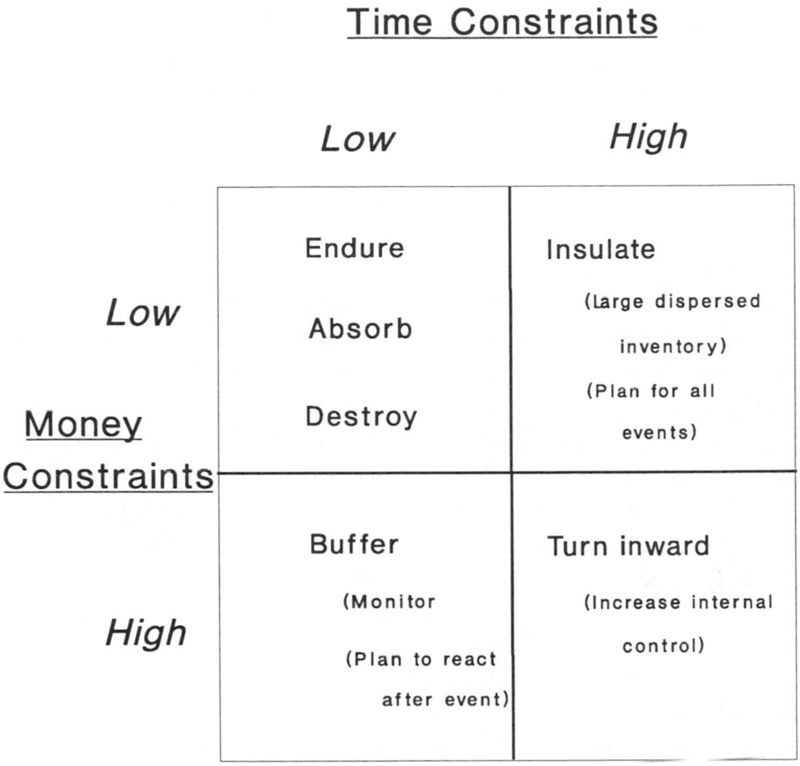

Time Constraints

	Low	*High*
Low **Money** **Constraints**	Endure Absorb Destroy	Insulate (Large dispersed inventory) (Plan for all events)
High	Buffer (Monitor) (Plan to react after event)	Turn inward (Increase internal control)

Taylor's scientific management theory of the early 1900s emphasized the potentially machine-like character of organizations. The popularity of the theory reinforced business firms' tendencies to view operations within the factory walls as separate from the society outside (McCurdy 1977:10–13; Shafritz and Whitbeck 1987:9–24). This concept of the organization as a closed system is prevalent today and encourages the tendency by managers to look inward for ways to offset external uncertainty.

Managers can either reduce the need to acquire and process knowledge, or increase the ability to do both (Galbraith 1977). Both involve greater controls on human behavior ranging from intensive socialization via training programs to strict uniformity in actual performance. To the extent possible, organizational nodes

[33]

are trained to react automatically to identified stimuli. Operations are analyzed and broken into discretely bound and often repetitive activities whose frequency and consistency tend to limit human variation. The length, format, and output of organizational operations are dictated closely. Managers enlarge monitoring and scheduling activities in order to regulate output and interactions. Individuals are placed in positions, or mechanisms are installed to verify compliance with the schedules or output of these activities. The goal is to make the organization more "clocklike" and free of deviations in performance.[20]

Aside from simply increasing in internal control, over the last two centuries organizations seeking certainty have turned to the perceived certainty of machine behavior in the mechanization of work. Today a powerful technological imperative operates among managers who exhibit a common and strong belief in the benefits of new, unknown technology. Not only is machine behavior perceived to be more predictable than human behavior, but newer machines are assumed to be better than equally adept humans. Tending to favor radical technical improvements over evolutionary improvement, managers are now replicating the nineteenth century's drive to factories in a rush to computers,[21] which are inserted widely and integrated in the control mechanisms of the organization.

As complex systems themselves, these new machines add to the rogue set and knowledge requirements of the organization overall. They tend to have unique information and precision demands that must be met if the equipment is to operate as planned, and they are apt to cost more and behave more unreliably than expected. Having assumed that nearly any new technology can be absorbed easily in the current arrangements in the organization, managers now find themselves responding to uncertainty coming from the new machines themselves. The effect is twofold.

[20]The literature on control in organizations is immense. For an introduction, see Perrow 1979, March 1965, and Shafritz and Ott 1987. For further reading, see Thompson 1967, Mintzberg 1979, and Morgan 1986.

[21]See, respectively, Ellul 1964 and Teich ed. 1977. For a good review of the historical development of technology, see Landes 1969.

Complex machines facilitate greater control by the managers and, when located in critical positions, then require yet more control of the uncertainty that comes with their internal complexity.[22]

The result is often nominally a more machinelike organization, more elaborate than planned. The additional concentrating, restricting, scheduling and/or monitoring change the structure of the organization, which itself becomes more internally intricate as the numbers, differentiation, and interdependence of the organization rise. Once-independent elements, sharing only a general source of resources, become more reciprocally linked, and depend on more nodes for more instructions, inputs, or outputs (Thompson 1967). The organization has grown more complex.[23]

The Positive Feedback Cycle

Complexity increases both the uncertainty and the desire for control. In a complex system with a large and unaccommodated set of unknown outcomes, knowledge needed for adaptation is less readily available. It is harder to predict which outcome will occur and whether it will be a rogue. Faced with a possible disruption in operations, managers need both time to react and accuracy in responses, in order to avert or mitigate the disruption. By its tendency to involve many nodes of the system very quickly, and by the number of nodes simply obscuring the originating sources of the disturbance, structural intricacy limits the speed and accuracy in responses to surprise.

A positive feedback cycle easily emerges in the presence of three conditions. First, the endemic scarcity of knowledge in complex systems means organizational uncertainty is particularly high. In response, managers of complex organizations are excep-

---

[22]See, respectively, Kaiser and White 1983, Augustine 1986, Spinney 1985, Kantrow 1980:7, Steele 1983:133–40, Winner 1977, and Posen 1984:55.

[23]The confusion about the nature of the interaction between the machines and their organizations extends beyond managers. A prominent military historian has written extensively about the problems of technology in war, without a theory of complexity to guide him. In his latest book on subject, he appears to have inexplicably decoupled the complexity of the machines and their logistics requirements (Van Creveld 1989:235).

[35]

tionally interested in control, and especially machines that promise to enhance control. Second, the integration of those machines in critical locations tends to introduce more uncertainty, reinforcing the desire for control.

There is no automatic check to this process if the organization normally has a great deal of difficulty testing hypotheses about control and, third, is highly constrained in time and resources. Furthermore, if this organization faces a great deal of uncertainty from the environment, the managerial bias toward control is exceptionally strong and amplifies the effects of the cycle. As this cycle runs its course, an organization can become significantly more complex and less predictable in nonobvious ways.

Complexity in the Military

Militaries are highly constrained organizations that have difficulty testing their hypotheses about performance. Although actual battle is a relatively rare event, the organization will face an extraordinary level of environmental uncertainty in wartime. The four options for reducing this external uncertainty—buffering, insulating, absorbing, and destroying—entail spies, allies, empires, and devastating battles, respectively. None of these options is easy, cheap, or quick.

Military managers are particularly prone to seeking internal control to compensate for external uncertainty. They alter equipment and organizational structures to acquire more knowledge about likely outcomes. From their inception, for example, military hierarchy and standardized drills were intended to increase the probability that units would perform predictably on the battlefield. The Romans forced new soldiers into the middle of a phalanx so the seasoned veterans surrounding them could decrease the possibility that the new soldiers would be able to flee. Even the much-discussed introduction of the stirrup was in essence an uncertainty-reducing device controlling the variations in the horse's likely behavior. Not only did the device reduce the variability of the horse, but it also had a small rogue set.[24]

Modern militaries seek complex technologies as a means of

[24]See, respectively, McNeill 1982:19 and Keegan 1977.

[36]

reducing uncertainty. History is replete with examples of technical advances that have changed the course of battles. Because fourteenth-century armor could be pierced by a bullet from a great distance, the development of the musket caused the mounted armored knights to disappear. The machine gun sealed the fate of horsed cavalry in World War I. The tank replaced the infantry as the dominant weapon in World War II.[25] It is therefore quite understandable that military organizations would be eagerly reaching for the latest technical developments.

To maximize correct responses to unforeseen events, a military organization particularly wants technical advances that offer more time or accuracy in responses. Not only is war chaotic, but the enemy watches, learns, and tries to make operations even more uncertain. For military organizations, then, the desire for complex machines whose salient characteristics are great speed and precision is almost overwhelming.

Just as in civilian organizations, these complex machines enter the military inventory with their own large set of rogue outcomes. The knowledge needed to accommodate these rogues is expensive. It is relatively scarce both in the organization itself and in the wider society—first because it is esoteric, and second because it is kept secret to deny the enemy information. A constrained organization acquires a costly knowledge burden when it acquires a complex machine.

Modern military organizations attempt to provide needed knowledge internally. In wartime, when information is needed it is needed immediately; waiting for an external organization to provide it is usually considered unacceptably risky.[26] For example, in terms of equipment uncertainty, most armies travel with inter-

[25]For an excellent review of these developments, see McNeill 1982 and O'Connell 1989. For a variety of knowledgeable and distinct perspectives, see Kaldor 1981, Mueller 1989, and Van Creveld 1989.

[26]During the late 1980s, ostensibly largely because of an imminent shortage of males in the conscript age-groups, Germany's *Bundeswehr* was developing a new army structure that would rely more heavily for key logistics services on civilian contracts in peacetime and, with a change in nomenclature, in wartime. The intent was to get legislation that at the outbreak of hostilities would automatically make those civilians into reservists subject to military command and keep them working. Although the legal difficulties with this plan were enormous, it is an example of the military deciding to rely on essentially outside organizations even in wartime (German interviews, 1987).

nal maintenance and supply units, and most have soldiers whose specialty is a particular kind of maintenance. These units constitute efforts to absorb the machine-related sources of external uncertainty and to subject them to internal controls.

The uncertainty of the machine and of the organization thus becomes linked in these maintenance units and in the effects that radiate into the wider population. For a military organization, the positive feedback cycle begins in these activities and then ripples through the rest of the organization. A given piece of equipment inexplicably seems to need more parts than anticipated, or it is nonworking more often than tolerable, and overall expenses for spare parts are much larger than planned. All these are recognized symptoms of increased uncertainty. Operating under constraints of time, money, or mobility, military managers set up organizations to check the flow and use of parts, monitor the use of the equipment, or install elaborate screens to prohibit parts usage that is not officially sanctioned. The U.S. Air Force, for example, has for years monitored the location of expensive subcomponents, "black boxes," for the F15/F16, the costs for this increased control, and its effects elsewhere on the organization, have never been included in the costs calculated for the aircraft itself, but they are undoubtedly high.

As the cycle develops, the more complex the weapons are in a military organization and the more scarce, critical, and unfunded the knowledge, the more complex the organization becomes and the larger the organization's rogue set. Given the nature of the environment and the requirements for certainty, increases in the complexity of a military organization increase the likelihood of poor performance in wartime.

A SET OF HYPOTHESES

A military service that has introduced a more complex essential piece of equipment into its critical operational elements makes a good case study of the interactions between the expensive unknowns of a complex weapon system and the adaptations or accommodations the using organization makes. In responding to

[38]

the rogue sets of complex machines, especially those for which knowledge is particularly expensive, the organization constrained in time and resources will normally show inadvertent increases in its own rogue set; it becomes itself more complex and more prone to disruptive surprises.⎤

Guiding the remainder of this discussion is a set of hypotheses, which depend in turn on two general organizational responses to unknowns which are widely observed in the literature.[27] First, the greater the set of unknowns (external or internal) perceived by the organization, the more the organization attempts to control events in advance. This effort entails acquiring knowledge about behaviors of humans or elements in order to reduce this uncertainty. Second, the more external (i.e., environmental) the unknowns *and/or* the more scarce (i.e., costly) the knowledge those unknowns require, the more intensively the organization's managers strive to control internal sources of uncertainty. When an organization's managers feel heightened threats from the environment, they attempt to ensure that those threats are not enhanced by unknowns from inside the organization. It is an understandable reaction to try to match a loss of control in one area with an increase of control in another. Preferable still is a reduction in overall organizational uncertainty.

From this discussion, the following set of hypotheses can be constructed:

> *Hypothesis 1:* The more complex and unique the machine, the greater its set of potentially deleterious unknowns (the rogue set) and the greater the resources or effort needed to obtain knowledge about potential outcomes to avoid an increase in organizational uncertainty.

> *Hypothesis 2:* The more critical the complex machines to organizational operations, and the more constrained the organization's resources or organizational latitude, the more internal control and spontaneous adaptation measures will increase and the more complex the organization will become in response.

---

[27]It is difficult to cite just one or two texts that demonstrate these responses when a wide variety of works do so. For good lists of the basic works on organization theory, see the bibliographies in Demchak 1987 and Scott 1981.

Hypothesis 3: The more complex a constrained organization and the more uncertain its environment, the less likely the organization will be able to respond rapidly and accurately to unpredictable environmental crises.

In other words, if a military seeks a complex weapon system to reduce its battlefield uncertainty, it then finds that the rogue outcomes of the machine require more knowledge than planned (Hypothesis 1). Organizations in general seek knowledge to accommodate their uncertainty and tend to use internal measures, especially controls, as substitutes for the knowledge needed to respond to a turbulent environment. Constrained by operational, resource, and theoretical limits, the military organization will be unable to acquire all the knowledge it needs to maintain the complex machine and will increase the internal controls to compensate. Those efforts, plus spontaneous informal adaptations, will produce a military that is more organizationally complex (Hypothesis 2). The increase in organizational complexity means the organization itself now has a larger set of possible rogue outcomes. Especially damaged is the speed with which the organization can accurately respond to unpredictable crises in a highly uncertain environment (Hypothesis 3). The military organization now knows relatively less about its own unknowns. The organization is poised for severe difficulties in both major and minor conflicts. For a constrained organization in a highly uncertain environment, complexity always has costs.

[3]

The Lure of the New Technologies

When the U.S. Army began its modernization program in the late 1970s, it was a greatly constrained organization. The Army had been attempting to modernize in the mid-1960s when the escalation of the Vietnam War intervened, drawing off the large budgets and political support necessary for broad technical advances (King 1972:188). Of the few new major weapon systems the Army was able to get to and through Congress in the period 1965–74, most were canceled or minimally funded. In particular, during this period, the Army lost a new helicopter project and two new tank projects to congressional cancellations.[1] By the mid-1970s the war was over, but the explosion in costs for updated equipment, training, and personnel put equally severe financial constraints on the organization. Accomplishing any modernization was likely to be both expensive and difficult.

This large, diverse post-Vietnam Army was also haunted by the apparent defeat of its technological advantage by a clearly technically inferior enemy. While the organization's leaders were willing to blame external actors, notably the press, for the strategic failures, the more technically advanced forces had nonetheless failed to provide stunning and decisive operational successes.[2] Draws and losses on the tactical level were far from ringing

---

[1] See Demchak 1987: Appendix A.

[2] The Vietnam experience spawned an extensive literature. For this point, see Gibson 1986:17 and Gabriel and Savage 1978:4.

endorsements of the higher-cost technology that the Army leadership was striving to make standard issue in the organization.

ATTEMPTS TO CONQUER UNCERTAINTY WITH MODERNIZATION

By the mid to late 1970s, the Army leadership, highly motivated to reduce battlefield uncertainty, renewed its interest in modernization. The new technologies were seen as the sign of superiority and invincibility. American technology, it could be said, had not been fully tested in the Vietnam War. The nature and costs of the war had not given the Army a chance to modernize fully. After Vietnam, the Army leadership was determined to be able to engage an enemy at all levels with American technology—and prevail. Strategically, modernization meant higher status both in the budget conflicts of the three military services and among other ground armies around the world. In short, the military could redeem itself by using technology to ensure a win next time (Gibson 1986:445–46).

A tank would lead the modernization: the M1 ABRAMS main battle tank. There were several reasons for the tank's return to prominence after several decades in which nuclear weapons, missiles, and guerrilla warfare dominated the thinking of planners. First, funding for forces dedicated to guerrilla war declined, and the Army leadership's attention focused on conventional missions. The European scenario had returned to the forefront (with the Vietnamese theater of war gone), and any battles there would certainly involve Soviet tanks.

Second, the Arab-Israeli wars of 1967 and 1973 featured the tank against Soviet equipment. This renewed emphasis on armor was spurred by the development and proliferation of effective man-portable antitank guided missiles. Technically advanced nations planning on using tanks on all sorts of battlefields, including nuclear ones, had to face the possibility that large and expensive weapon systems might be destroyed by relatively cheap missiles.[3]

---

[3]This opinion is not universal. Weigley (1984:587) asserts that precision-guided munitions (PGMs) represent the new tactical revolution, but their cost inhibits their use, and hence revolutionary changes are also limited.

[42]

As more nations equipped their armies with these missiles, interest in increased armor protection, especially the new composite armor, emerged along with increased efforts to upgrade the entire weapon system. The two developments tended to balance one another, and the tank returned to its premier position in army inventories.

Third, successful tactics borrowed from the Israelis stimulated the interest of U.S. doctrinal planners in the possible ways to improve both mobility and firepower, two of the three basic elements in tank design. Tactical success as defined by U.S. doctrine came to rely more heavily on tanks that had a long effective range in conventional firepower coupled with infantry fighting from very mobile armored personnel carriers and helicopters carrying missiles and more troops (FM100-5 1976).

Fourth, the tank's other main competitors—missiles and helicopters—failed to provide the tactical success and resilience, respectively, of the tank.[4] Used as longer-range and more accurate artillery, missiles had not proven to be the tactical breakthrough heralded in the 1960s. Of the current weapon systems, only helicopters constituted an alternative to the tank on a major battlefield.[5] However, the machine had not provided success in Vietnam, and given the greater vulnerability of helicopters to common battlefield hazards, such as small-arms fire, the problems with attrition would be enormously exacerbated.[6] Exchang-

---

[4]An armor specialist has observed that the improvements of the armor have advanced so much that the antitank weapon fired by the individual soldier is more in danger of becoming obsolete than the tank. This thought is echoed cautiously by Middleton (1985:112); see also a commentary in *Armed Forces Journal* 1984:91.

[5]One of the newer military thinkers in the Army, Brigadier General Huba Wass de Czege, has said that the current use of the helicopter is like the use of the tank in the U.S. Army prior to World War II, i.e., helicopters are dispersed in support. He claims the next major change in tactics will emerge when appropriately armored helicopters are consolidated and used in mass assaults like tanks (Wass de Czege 1986). The recent recognition of aviation as a separate branch of the Army with a separate school and development center is an essential first step in developing the organizational consensus for a doctrinal change supporting the helicopter. Whether a change emphasizing the use of helicopters like tanks would constitute such a novelty in tactical doctrine is open to question.

[6]High and unexpected losses of helicopters due to small-arms fire plagued operations in the 1983 invasion of Grenada as well (Gabriel 1985:180, Adkin 1989:356–57).

ing the tank for a helicopter would have been financially catastrophic.[7]

Finally, the specter of Soviet tank superiority in Europe was less obscured by the Vietnam War, and a new tank was strategically desirable, if only to dispel any Soviet thoughts of adventurism on Europe's North Plain. In addition, Congress was willing to pay for a new tank. It had penalized the service for poor performance on two previous tank projects and under certain conditions was willing to see the Army redeem itself with a third try.[8]

Hence, when the Army's leaders sought to reduce the organization's battlefield uncertainty through modernization, they placed a new tank in the lead. They introduced into their already complex organization a key weapon system that was not intended to be, but was, more complex than its predecessor. Thus, two complex systems met, each dependent on the other for successful operations.

DESIGN, DEVELOPMENT, AND SURPRISE

Having led the Army's modernization program, the tank is an excellent case for studying how a critical weapon system's complexity affects the organization that absorbs it. The M1 ABRAMS main battle tank is more complex than its predecessor, the M60 tank, and its introduction sparked the changes producing a more complex tactical organization.

The design and development of a system have enormous impli-

---

[7]In 1982 the Army planned to acquire 5,096 M1 tanks from fiscal year 1983 to fiscal year 1987 for $13.3 billion. The same amount of money would buy one-fifth as many attack helicopters at $12.9 million each (Slatkin 1982:xiii).

[8]The two tanks were the MBT-70 (a joint U.S.-FRG project canceled in 1970) and the XM-801 (a U.S.-only follow-on project canceled in 1972). In both cases Army representatives were unable to produce a working tank for cost figures even close to their promises. In the case of the XM-801 in particular, Army representatives were caught lying to Congress. As a result of what seemed to be a generally poor level of competence, Congress took relatively drastic measures by canceling the tanks and putting severe restrictions on the new tank project. See Demchak (1987: Appendix A) for a discussion of how the Army lost much organizational autonomy to Congress as a function of the Army leadership's misunderstanding of the pitfalls of technological development.

cations for the system's set of potential surprises. Some designs make acquisition of scarce knowledge about a system relatively easy, others do not. Similarly, some developmental processes may be better than others at providing data about rogue outcomes before they disrupt operations.

In the defense community, and especially in procurement, much is made of the differences between evolutionary or radical development of weapons or equipment. Evolutionary development occurs over a long time with many models in a series; radical development emerges with the machine designed as a whole, unrelated to existing machines. The continuing debate over quality versus quantity in weapon systems, begun in the 1970s, frequently takes sides along this distinction. According to the quantity side, evolutionary equipment is more reliable and in great numbers more capable; their opponents say that quality is individually assured in radically developed equipment because it incorporates the latest discoveries and that any unreliability problems will smooth out over time (Clark et al. 1984).

It is a mistake to make too much of this distinction. The knowledge available to meet surprises is what counts, and evolutionary development can only help. Evolutionary development tends to accumulate information about rogue outcomes automatically over time, whereas a radically developed machine arrives with most of its outcomes unknown. Conversely, a very complex system can be twenty years old and impose a greater knowledge burden on the organization than the brand-new but much less complex radically developed machine. In practice newer weapon systems are usually much more complex. It is the combination of radically developed and more complex that imposes the heaviest knowledge burden on the organization.

When the M1 was introduced it was displacing an evolutionary tank, the M60, developed over twenty years. Intended from its inception to be a highly integrated system,[9] the M1 was from the start both radically developed and complex. The history of each

---

[9]The series of political events producing the contract for the M1 ABRAMS tank, as well as the conditions of that tank's development, are discussed in detail by Demchak (1987: Appendix A).

tank's development represents an interesting comparison of the potential rogue outcomes built into each.[10]

M60 Main Battle Tank Program

The advantage of evolutionary development is not that errors and mistakes will be eliminated, but that iterations tend to automatically reveal or accommodate most of these before the tank is issued to the forces in the field. If the problems are not identified before the field gets the tank, similarity to the older design helps the tactical forces accommodate the new unknowns. For example, when the M60 was initially issued to the field in 1960, there was a severe problem with spare parts and ammunition for the new tank.[11] The M60's commonality with previous tanks, however, helped make the necessary knowledge less unique and disruptive.

Another advantage of an evolutionary development of a tank is that Army personnel and their experiences in the field are more deeply involved at the decisive points. Commanders in the field and budget-makers in the headquarters automatically enter input, because each iteration requires gathering data from the field and consulting those who make funding decisions.

The M60 began, however, with a big advantage: it was not a very complex tank, and each round of modifications did not require extensive redesign of the entire system. In 1949, as often happens in postwar periods, the Army did not get the funding for its proposed three-tank system. Finally the Army up-gunned, up-engined, and placed a new turret on the M26 Pershing tank, the last heavy tank (46 tons) built during World War II (Green et al. 1955). Relabeled a medium tank, the altered M26 was designated the new medium tank M46.[12] The M60A3 tank evolved from

---

[10]Material drawn primarily from a report done for the Assistant Secretary of Defense for Manpower, Reserve Affairs, and Logistics by Logistics Management Institute, LMI report ML 101, written by Lucas Bragg and Chris Demchak in 1982.

[11]"The introduction of the [M60A1] tank into widespread use had revealed the usual need for numerous modifications to improve its performance and reliability" (Hunnicutt 1984:199,165).

[12]The "heavy tank" nomenclature has not been used again, even though the current M1 is heavier than the older "heavy" M26 tank.

the M46 tank through a series of product improvements over twenty years.

In 1950 the M47 was designated from the developmental M46A1; the new tank had a M46 hull with the T42 turret and a new 90 mm gun. In 1952 the M47 in turn became the M48 with the addition of a new turret. In 1955, fuel injection was added, to produce the M48A2 tank. In the same year, a diesel engine was added for increased range, reliability, and safety, to produce the M48A3. In 1956, new power packs were added to the M48A3 to create the M48A2E1 and directly pave the way for the next experimental tank, the XM60 (Hunnicutt 1984, Bragg and Demchak 1982).

With the new British 105 mm barrel and U.S. breech, the XM60 was designated the M60 and classified as standard in 1959. As an evolutionary system, it shared a high proportion of its component parts with its predecessors, so its introduction was not as much of a disruption for the organization as the later M1 would be. In 1959 the M60 had 50 percent of its parts in common with the M48A2, 25 percent with the M48A1, and 20 percent with the basic M48 (Bragg and Demchak 1982).[13] In addition, during the testing, several reliability requirements were waived in order to achieve this classification. The evolutionary nature of the tank, however, mitigated the worst possible effects of this incomplete testing. Much was already known about the tank through experience with its predecessors.[14] In 1962 the M60's turret was modified to

[13]The high proportion of common parts has meant that stockages have been able to evolve slowly to accommodate the newer versions without dramatic disruptions in the organization. This long experience supported a widespread presumption that the disruptions of new tanks were at best temporary and that the organization would easily adjust. The organization's experiences with the M60A2 missile-firing variant of the M60 were forgotten.

[14]This official disregard for reliability requirements would have been unprecedented before World War II and only acceptable during the war for matters of extreme urgency (Green et al. 1955). The fifteen-year effort to get a new tank designed and funded produced within the managerial level of the Army a peacetime urgency and willingness to accept defects in such nonobvious areas as reliability. This pattern becomes more common with time, as costs, delays, and fear of Soviet technical dominance grow in Army weapon system development. For other examples, see Demchak 1987.

[47]

produce the M60A1, the standard tank until the early 1980s.[15]

In 1978 the first M60A3 was produced from a modification of the M60A1. Key improvements include a reliability-improved (RISE) engine, laser range-finder, built-in stabilization, and a new solid-state computer. In 1980 the M60A3TTS was produced, an M60A3 with the new tank thermal sight (TTS). The M60A3 is the most current and said to be the last version of the M60 in the Army inventory.

The M1 ABRAMS Tank Program

The purpose of the M1 ABRAMS development was to improve dramatically the performance of the U.S. main battle tank. The new tank was intended to incorporate state-of-the-art technical advancements without undue costs or complexity. The key improvements in the M1 design of the tank were incorporation of a hull and turret constructed of special armor, upgraded armament with an increased rate of fire and stowed load, and a 1,500 horsepower turbine engine (LEA 1979). Another key improvement was the new stabilization system, which made it possible to fire on the move.[16]

Under a cloud from the Army's previous tank development efforts, the XM1 tank program was the Army's effort to prove to Congress that the Army could wisely buy, develop, and use advanced technology in critical weapon systems.[17] Initially established in December 1971, the final XM1 program was approved by the deputy secretary of defense in January 1973. In June 1973, each of two contractors agreed to develop an XM1 prototype tank

---

[15]An exception to the evolutionary development of the M60 was the M60A2. During the 1960s, the Army led the nation in computer development and zeal for missile weapon systems. In 1966, a variant of the M60A1—the M60A2—emerged with a new complex missile-firing turret. Production began in 1966, but because there were significant mechanical problems these tanks were not issued to the units until 1972. After nine years of reliability problems, all the remaining M60A2s were removed from Europe in 1981. The tank model was discontinued, its failure ascribed to putting a missile on a tank, not to complexity itself, and the lesson on complex systems was lost.

[16]Most tanks must stop to fire accurately.

[17]For a discussion of the difficulties the Army had in obtaining congressional approval for new tank programs, see Demchak 1987: Appendix A.

and some test rigs that met the military requirements but remained within an average hardware cost of $507,790 (1972 dollars) per unit. In November 1976, Chrysler was selected to complete the tank (Schreier 1977, Kane 1981).

Unlike the evolution of the M60, Army personnel were not deeply involved in the design evolution of the new tank. They contributed mainly in relatively high-level monitoring of the program and the later testing of prototypes. In tests conducted in 1976, highly trained contractor personnel performed all maintenance tasks and controlled all tools and test equipment. The sample size and test mileage were low. In later tests, conducted in 1978 for the sake of the schedule, normal Army maintenance procedures were neglected; many contractor personnel performed tasks intended for one soldier. Results were poor and the assessments of maintainability, durability, and reliability were considered "soft" (Kane 1981:4–36, Bragg and Demchak 1982).

These results were all the more problematic because Chrysler was to rely on the "smart machine, dumb maintainer" maintenance philosophy in constructing maintenance support equipment as well as technical manuals, and tools (collectively referred to as TMDE). Only procedural remove-and-replace actions were to be performed, and at the lowest levels (called "organizational" or "unit" levels), and minor component repairs were to be done at the next stop, the direct-support (DS) level. The bulk of Army in-house skill with the M1 was to be located initially at the depot and later in the rear of the theater, at the general-support (GS) level, and only for major repairs, such as overhauls. It was presumed that the maintainers in the forward areas would need very little knowledge about the intricacies of the tank.

The remove-and-replace maintenance concept depended on a highly reliable weapon system, but diagnostics continued to present major difficulties during tests. To keep assemblies and subassemblies functioning, major subcontractors developed a large and unmanageable array of contractor test equipment. And the test equipment built into the tank was proving to be inadequate. In 1979, the Army gave Chrysler a supplemental contract to produce one piece of test equipment for all major subsystems of the tank—the Simplified Test Equipment M1 (STE-M1). It failed,

[49]

however, to find the source of the malfunction 74 percent of the time, and the electronic test set intended for direct-support (DSETS) 38 percent of the time. The diagnostic problem persisted.

The tank was displaying unexpected surprises. Out of thirty-four problems analyzed, for 34 percent to 56 percent of the time contractor assistance was required, the wrong fault was identified, and good parts were unnecessarily removed. Even the backup—alternate troubleshooting procedures and technical manuals—could not diagnose the problems (LEA 1979, Kane 1981). Major problems remained after the tank was officially accepted by the Army (LEA 1979).[18]

The first tank was completed in February 1980. Routine production began at the Chrysler-operated (now General Dynamics–owned) Detroit tank plant in 1982, stabilizing in 1984 at about sixty tanks a month. The final inventory is anticipated to be about 3,500 M1 tanks and a slightly greater number of the follow-on tank, the M1A1 (Binkin 1986:26, Schreier 1977:460).

Assumptions about the tank's knowledge burden, not the demonstrations in the testing process, governed the decision to accept the new tank. These assumptions also were integral to the success of the new remove-and-replace maintenance philosophy, which justified lower training levels in maintainers. Although the M1 reportedly had statistically met or exceeded all reliability requirements, it did not do so without alterations. Defense Department program managers expect reliability improvements to occur and routinely adjust test data under broad assumptions of future improvements.[19] For example, a key figure for a tank is its mean mile between system failures. For the M1 the original requirement was 101 miles and the actual test figure was 75 miles, but the

---

[18]The Army Systems Acquisition Review Council (ASARC) is the last step in the Army's acquisition decisions before Congress and the secretary of defense make funding decisions.

[19]Final test results for Defense Department tests are altered by certain assumptions about future corrections of problems uncovered during the tests. For example, quality control has been responsible for 30 percent of all failures, so, all things being equal, future improvements in production quality control would make the tank more reliable than the tests suggest. Availability was to fall between 89 percent and 92 percent. If the test results were significantly lower than that—and they were—it was considered acceptable to adjust the results to reflect what availability would be with few quality control errors. Generally only the adjusted figures, not the raw data, are announced (LEA 1979).

announced adjusted figure was 98 miles (LEA 1979). The experiences in the deployed units indicated that the unadjusted test figure was more likely to be correct. In 1983, four years after the start of production and two years after the M1 ABRAMS' issuance to the field, the tank averaged 67 miles between system failures (Binkin 1986:64).

When such assumptions are wrong, the costs to resolve the problems after the equipment is in the field are often greater than if the problem had been resolved during the development process (Kaiser and White 1983).[20] Seven years after the tank was introduced in the field, it continued to show significant diagnostic problems, and costly special training had to be hastily reintroduced to provide scarce knowledge.[21] In the field, the tank was often surprising.

COMPARING COMPLEXITY

Hypothesis 1: The more complex and unique the machine, the greater its set of potentially deleterious unknowns (the rogue set) and the greater the resources or effort needed to obtain knowledge about potential outcomes to avoid an increase in organizational uncertainty.

How capabilities and components are ordered and linked in a tank determines the machine's complexity. Armored vehicles are compromises between four competing requirements: firepower, maneuverability, speed, and crew protection (Green et al. 1955:283). The mission of a tank is generally so broad that the order of priorities for capabilities is often not clear. For example, the M60A3's Material Fielding Plan offers a generic mission applicable to all U.S. Army tanks. The plan addresses such requirements as cross-country mobility, an extended cruising range, crew ballistic

---

[20]For a typical five-day exercise in the field in Europe in late 1982, some 45 to 50 percent of an armor battalion's fifty-eight tanks were not-mission-capable (NMC) by the end of the exercise, a figure nearly the same for the 1981 test results in a six-day exercise (Demchak 1983).

[21]The Army reported that its European tactical units were showing a "significant problem with the M1 and the M2/3 [Bradley...fighting vehicle] maintenance in the area of diagnostic troubleshooting" (Tice 1988:6).

and nuclear protection, night operations and nearly all temperature operations, three- to four-foot river and trench fording, surmounting three-foot vertical obstacles or 30 to 60 percent grades, and of course defeating enemy armored vehicles (Bragg and Demchak 1982). It is important to note that there is no discussion of reliability, battlefield repair, or effects of this equipment on the rest of the organization.

Without a better understanding of the costs of complexity to help guide decisions about making trade-offs among the various requirements, the inadvertent production of a highly complex tank occurs easily. A wide variety of components promises a wide variety of capabilities, which are needed for a diverse set of missions. Furthermore, modifications to an initial design can disrupt the pyramid of compromises that went into the entire design or prototype; successive development decisions are often fixes to previous fixes, irrespective of the contractor or of other specialized knowledge involved.

Because a lack of knowledge is symptomatic of a more complex system, the relative difficulty of diagnosing the sources of failure is a common and useful surrogate indicator of complexity. The more complex the system, the harder it is to know for sure which set of components has failed when something does not work correctly. The electronics industry commonly uses "failure mode analysis" to identify and statistically assess potential diagnostic difficulties.[22] In this process, the universe of possible failures is identified, and starting with the most obvious each identified failure is traced back to all possible sources of that failure in the system. This tracking involves counting nodes and connectors for each node and assigning probabilities of failure to all.[23]

In electronic systems, failures are particularly problematical. Because of the large rogue set of possible sources of failure, electronic systems fail in a random pattern, unlike the gun or

[22]In electronic systems prone to unpredictable failures, the key to effective maintenance is accurate identification of the source of the problem. If diagnosis of a failure does not lead to the correct repair, then the only repair procedure remaining is expensive replacement of all suspect parts.

[23]Similar to the NDI analysis described in Chapter 2, failure mode analysis directly assesses numbers and interdependence while indirectly assessing differentiation.

engine systems. This makes diagnosis of the problem particularly difficult. For example, if a system has numerous interconnected nodes, any failure has a greater chance of involving a greater number of nodes and/or having multiple original sources than would a less integrated system. And a multiple source failure is more likely in a more integrated system. The more complex the system, the more integrated it is, the more likely it is to have multiple source failures,[24] and the more difficult failure mode analysis will be.[25] The time required to identify the multiple sources of failure increases enormously.[26]

In M1–M60A3 research performed, data from failure mode analysis was used to construct a "coefficient of complexity."[27] This coefficient combined a numerical calculation of the universe of potential failures and the amount of information needed to diagnose a failure, with another numerical calculation of the amount of information available in the system to use in diagnosis. The larger the first calculation (the suspect set), the larger the second calculation (the sensor set) needed to be in order to provide the necessary diagnostic success.[28] A large suspect set means a poten-

[24]"Many of the cables in a combat vehicle are branched, as opposed to point-to-point. Consequently, some of the cable connectors contain up to 61 pins and the other end of a particular circuit may show up in one of several branches of the cable" (Sarna 1982:17).

[25]Research by Wohl (1980) indicates that, for electronic systems of average complexity, 90 percent of the failures can be diagnosed in two hours. The remaining 10 percent can take anywhere from two hours to infinity.

[26]In principle, one cannot identify every possible outcome, because for every system there is a set of unknowable unknowns. With simple systems this set may be small and irrelevant, but with complex systems the universe of unknown outcomes is so large that identification of the outcomes that *can* be known is often very expensive and time-consuming. Hence, identification generally continues to the extent permitted by available funds and time, not to some broadly accepted level of certainty. Depending on the amount of research time and money invested, the electronics firm selects which failures to include or document in troubleshooting manuals and/or automatic test equipment, and which to exclude or not pursue.

[27]In 1979 a team at a nonprofit defense think tank, Logistics Management Institute (LMI), began research on using engineering principles to compare the complexity of the M1 tank with that of the M60A3 tank. Two LMI engineers, Lucas Bragg and Franz Nauta, devised the mechanisms for assessing machine complexity later applied by Bragg and myself (Bragg and Demchak 1982).

[28]The suspect set is derived from calculation of the average number of components associated with a failure and actual or proxy failure paths, using surrogates such as active pins and cables and connections. The suspect set can be minimized by reducing connectors and possible inputs, that is, reducing the integration of the

[53]

tially large set of rogue outcomes, and the sensor set represents available knowledge or ability to acquire information when needed.

For example, if the test equipment in the sensor set does not provide enough information, as demonstrated in Figure 3, a mechanic able to diagnose only at the black box level would use the sensor set available and inaccurately replace both boxes C and D, taking a good box, D, out of service. Several sources of failure would not even be identified by the test equipment because they are unpredictable and therefore unlikely to be included in the design considerations of test equipment.

Ultimately, the result of such calculations indicates whether the actual sources of a given failure will be even more unclear or ambiguous after all the sensor equipment is applied. In a more complex system, the index of complexity is an index of ambiguity or uncertainty. A more complex system is one in which, for failure symptoms z, failures x and y are either more likely to occur simultaneously or, having occurred, less likely to be accurately identified. In other words, the average ambiguity group for a given failure is the set of suspect sources divided by the sensors available and will be greater for a complex system. It is simply more difficult to know what went wrong when the machine fails.

By these measures, then, the M1 ABRAMS tank is a more complex tank than the M60A3. In tests comparing three subsystems common to both tanks[29] and essential for tank performance,[30] the M1 has a larger suspect set overall (Table 1). In the engine and the stabilization system the M1 systems were more complex than the M60A3 systems, ranging in ratio from 2:1 in the stabilization system to an average of 3.6:1 in the engine. The third subsystem, the laser range-finder, is less complex on the M1 than on the M60A3 retrofitted, with a ratio of 1:1.6 in the suspect set (Nauta

---

system, but not necessarily by using partitioning. The sensor set is measured by how complete the information on all possible failures is and by how specifically it can accurately identify the sources of a failure.

[29]See the Addendum at the end of this chapter for a discussion of these three systems.

[30]In one unit in 1982, the engine, turret, and fire control systems (of which the laser range finder is a key part) were responsible for 70 percent of all the major tank failures (Demchak 1983).

Figure 3. Multiple source failures and misdiagnosis of sources of failure symptoms

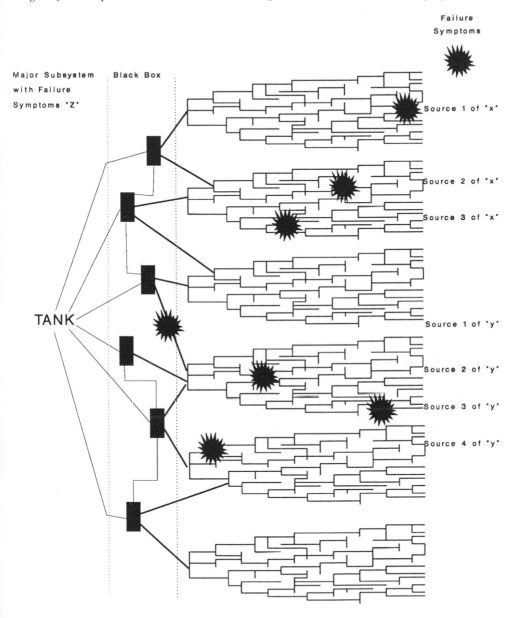

and Bragg 1979). Consistent with the presence of a larger rogue set, knowledge requirements are commensurately larger (Bragg and Demchak 1982).

Table 1. Average ambiguity groups[a] for M1 and M60A3 tank subsystems

	M1	M60A3	Ratio M1:M60A3
Engine system			
Abort/stall	12.0	6.0	2.0 : 1.0
No start	40.0	10.0	4.0 : 1.0
Laser range-finder system	3.3	5.6	1.0 : 1.6
Stabilization system	15.0	4.2	3.6 : 1.0
Overall ratio			2.3 : 1.0

Source: Bragg and Demchak 1982:3–7, 3–8.
[a]Average ambiguity group = (suspect set / sensor set).

Greater knowledge requirements are also indicated by the actual physical size of the ambiguity group; for this set, only figures for the M1 were available. Of the 544 failures identified for the stabilization system, only one-third were testable by the required specialized test set when the tank completed its final operational tests, and 40 percent of the identified engine failures could be diagnosed using that equipment (Bragg and Demchak 1982). The M60A3 can be diagnosed with a simple multimeter, but the M1 can be diagnosed without the specialized M1 test equipment only with great difficulty (Nauta 1983:4–19). Because the failures identified were single-source failures, not the multiple-source failures common on complex equipment, unmet knowledge requirements reside in the unknown number of multiple-source failures and in the 60 to 70 percent of failures not diagnosable by specialized test equipment.

An increase in knowledge requirements is also indicated by the increase in the number of types of specialized test equipment required. The additional electronic sophistication has added to the support burden of the M60A3 tank as well. Some of the special

[56]

test equipment required to support these tanks are a computer cable test set, computer field test set, laser range-finder test set (with hot mock-up of components), turret electrical test set, stabilization test set, and receiver/transmitter tester (used with laser test set) (Bragg and Demchak 1982). Each piece of test equipment has its own failure patterns and limitations, which add to the uncertainty of the organization.[31]

Certain additional items cannot be maintained easily in the field—for example, the laser range-finder must be repaired primarily at depot level. Of the 781 repair parts associated with the laser, only 2.4 percent can be replaced at the organizational level, and only 12.7 percent at the DS or GS level (LEA 1979). With more than 85 percent of the parts of the laser range-finder required to be returned to the depot for repair, the laser range-finder's availability depends heavily on the survivability of air deliveries or a supply of extra systems in the theater. Without this system, the M1's accuracy in destroying targets is much reduced.

The design of the M1 makes the tank more knowledge-intensive and dependent on external equipment than the M60A3. This complexity means the addition of equipment, tools, manuals, and procedures. In terms of the equipment inventory, then, numbers, differentiation, and interdependence increases have been imposed on the Army system. Furthermore, increases in complexity induce increases in diagnostic time that are lognormal rather than linear in nature (Bragg and Demchak 1982). With an increase in tank complexity, users need to know more, and it takes longer to find out what they need to know.

Effect of Complexity on the Knowledge Burden

The new tank, the M1 ABRAMS, is a more complex tank than its predecessor, the M60A3 tank. The M1 failure mode analysis

---

[31]These increases are particularly onerous when even skilled technicians find it difficult to determine where to hook the test equipment up to the tank: the seven footlockers make it bulky to use, its diagnostic effectiveness is limited because it isolates to an ambiguity group of only three boxes, and it is too slow because the circuitry must cool down or the test results will not be accurate (Nauta 1983).

data showed increases internally along at least two of the three indicators of complexity. Both numbers and interdependence rose. With the inclusion of the external requirements for test equipment and types of parts, differentiation—the third indicator—also rose. Other key battle systems—the Bradley armed personnel carrier, the MLRS artillery system, and the Apache attack helicopter— seem to display the same increases in internal complexity.[32]

These new machines impose a higher knowledge burden on the organization than did their predecessors, and the required knowledge is scarce and costly (see Figure 4). Initial estimates of the M1's reliability proved inadequate, indicating insufficient knowledge; diagnostics proved to be a major difficulty that continued long after the tank was in use. In the development of the M1, obtaining this reliability and maintainability knowledge within established time and cost schedules was considered too costly; multiple and highly skilled contractors were concentrated on tasks intended to be performed by a single soldier. In operational tests, test equipment consistently performed poorly, requiring expensive upgrades to improve performance; alternate test procedures also were inadequate, indicating inadequate knowledge of the system by the designers of the test procedures and equipment.

The original 1975 estimate of the cost of 3,325 M1 tanks was $4,900 million, but by 1978 the total costs had increased nearly 50 percent, to $7,251 million.[33] Originally intended to cost less than $1.2 million each by production in 1979,[34] the M1 tank cost $2.7 million by the early 1980s and $3 million by 1988 (Kane 1981:4–8, Rasor 1983:34). Similar increases have occurred in the procurement costs of other key weapon systems from 1970 to 1982 (in 1983 constant dollars): a squad helicopter from $1.0 million to $7.6 million; an armored helicopter from $3.5 to $15.0 million; and the armored personnel carrier from $200,000 to $1.5 million. (Binkin 1986:26, 44).

[32]See Demchak 1987: Chap. 3, add. A, for a review of other systems' complexity. Demchak 1987 (chap. 7, add. A) presents the effects of these and other new systems on differentiation in the technical core.

[33]In constant dollars, to be paid by 1987.

[34]The 1979 cost was calculated by assuming an average 10 percent inflation for the period 1972–79 and applying this inflation index to the original mandated price of $507,790 per production tank (Kane 1981:4–8).

Figure 4. Additional knowledge requirements of the M1

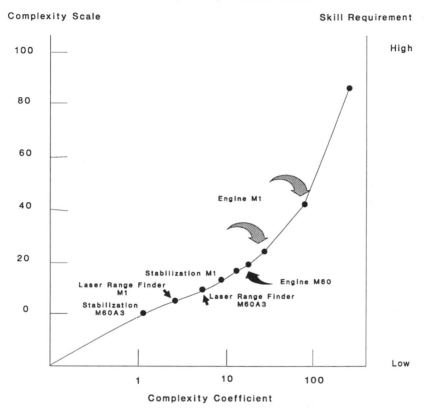

Scaled Complexity Coefficients

Complexity Scale   Skill Requirement

In the late 1970s and early 1980s, congressional debates focused on the extension of a contract to a second supplier of M1 parts, to increase the speed of delivery but primarily to control the costs of the knowledge associated with the production of the tank. This increase in internal complexity and the associated increase in money and skills needed to accomplish original plans provide support for Hypothesis 1; such complex machines impose a knowledge burden that is both difficult and expensive to accommodate.

[59]

APPENDIX TO CHAPTER 3: CRITICAL SUBSYSTEMS
OF THE M1 AND THE M60A3

Both the M1 tank and the M60A3 tanks are fully tracked and heavily armored land combat vehicles operated by four-man crews consisting of a commander, gunner, loader, and driver. The main weapon of each tank is a 105 mm gun mounted in a traversable turret. (According to original production plans, a 120 mm gun has replaced the 105 mm in the M1A1 version currently under production.)

Both tanks have a laser range-finder (LRF) that determines target range by transmitting a pulse of laser light aimed at a target and converting the time from transmission to reception into range. In the M1 this range-finder is fully integrated into the gunner's primary sight (GPS); in the M60A3 the range-finder is a separate system.

The M1 tank is powered by a 1,500 hp gas turbine engine driving an automatic transmission that has four forward and two reverse gears. With governed maximum engine speed, the vehicle's top speed is 45 mph (60 mph if not governed) on level hard-surface roads and 30 mph on cross-country terrain. The M1 averages about 5 gallons of fuel per mile. Its planned operational range was 300 miles, but its actual operational range has been much lower, 131 miles (Kane 1981:4–12, Rasor 1983:36).

The M1's fire control system is continuously monitored and controlled by a digital ballistics computer that also performs built-in test functions. The fire control system consists of the gun/turret drive and stabilization system, the ballistics computer, the gunner's primary sight (GPS), the laser range-finder, the commander's weapon station sight, and the muzzle reference sensor. The gun/turret drive and stabilization system provide line-of-sight stabilization in elevation and gun/turret stabilization in azimuth (Bragg and Demchak 1982, Schreier 1977).

The M60A3 is powered by a 750 hp turbocharged compression-ignition diesel engine. The vehicle's top speed on hard-surface roads is 30 mph, and 12 mph cross-country. The M60A3 averages about 3 gallons of fuel per mile. Its technical operational range is

310 miles, but its fielded range is 160 miles (Kane 1981:4–12, Rasor 1983:36).

Less integrated than that of the M1, the M60A3's fire control system consists of either an M35E1 gunner's periscope and a M21 reticle projector unit or tank thermal sight (TTS) An/VSG-2 mounted at the gunner's station, an M36E1 commander's periscope, a laser range-finder, and a ballistics computer system (the M35E1 and M36E1 amplify ambient light, whereas the TTS An/ VSG-2 detects thermal energy). The computer calculates data required for aiming the main gun, tracking, and firing at moving or stationary targets, and it regulates power and most of the signal processing, computing, and built-in test electronics of the fire control system. The turret stabilization system allows simultaneous tracking of the target and firing control of the main weapon while the vehicle is moving (Bragg and Demchak 1982, Schreier 1977).

[4]

The Army as an Organization

With the M1 ABRAMS a complex tank entered the inventory of a constrained organization that was consciously seeking minimal levels of uncertainty. The U.S. Army was already a complex organization; it had elaborate personnel management and equipment support rules developed over the previous half century.[1] Several times the Army had attempted to simplify its policies with new division structures, new personnel specialities, and standardization of equipment requirements and issue, but each time the simplification was relatively short-lived (Weigley 1984).

This chapter lays the groundwork for the next two chapters. To clearly delineate key relations within the Army structure, the organization is divided into three parts:[2] the institutional level, the managerial level, and the technical core level (Thompson 1967). These provide a way to divide the organization's function analytically in order to trace the effects of complex equipment on the tactical Army. Chapter 5 focuses on the managers' decisions in

[1] In 1983 the organization guided its behavior with about 1,200 Army regulations (ARs), more than 1,700 field manuals (FMs), and more than 16,000 technical manuals (TMs) (U.S. Army, DA PAM 310-1, 1983).

[2] The term "level" used in discussions of structure means groups arranged at successive intervals, usually in a hierarchy. Unfortunately, the source documents for this research use the term widely for different types of activities: organizational level, maintenance level, level of repair, support level, etc. To minimize the confusion, I use "level" in two distinct cases: referring explicitly to J. D. Thompson's levels of organization and, unavoidably, when discussing the Army's structure of successive maintenance functions—that is, the organizational, direct support, general support, and depot "levels." To describe the internal structure of the technical core, I use the Army synonym for level: "echelon."

the managerial level, Chapter 6 on the responses of the technical core—the tactical units.

STRUCTURE

The institutional level is normally the guiding element of an organization, and in the Army its activities cover an enormous spectrum. These range from testifying before Congress, making decisions on the proper structure for an armor division reduced in force to 10,000 troops from the normal 16,000 troops, and approving policies that change equipment, structures, and recruiting policies.

This level consists of the headquarters of the Department of the Army (HQDA), the secretary of the Army, the chief of staff, major theater and commodity commanders, and the direct staff of the secretary or chief. Composed primarily of general officers and their direct staffs, this level is managed by a dominant (leadership) coalition guided by an "inner circle" (Thompson 1967:130–40). Each chief of staff of the Army—the top military position—gathers a group of fellow generals to plan and implement the ideas the new chief carries into the job. Senior officers who stay in key positions after the new chief is ensconced may be considered part of the inner circle. The Army's dominant coalition numbers about twenty-five generals in influential command or staff positions; the inner circle is made up of less than ten.[3]

The managerial level is the intermediate level that translates the guidance of the institutional level and the needs of the technology of the organization into rules and decisions for the technical core. In the Army, aside from general officers who are senior commanders, this level consists of the theater-level staffs and the staff of the commodity commands, a total of eighty-five support commands and fifteen major commands (AUSA 1986:288, 352). By

---

[3]Numbers are estimated by me from the AUSA (1986:286–352). Some may not be in the chief's camp, but their influence in Congress or the White House makes them difficult to displace. Despite the universalistic tenor of the organization, as Janowitz observed, "career experiences and *personal alliances*...are at the heart of...differing concepts of military leaders" (Janowitz 1971:16; italics added).

[63]

historical precedence, the support commands are given the status of combat commands. Field combat commanders commonly rotate through these commands en route to top positions in the Army's headquarters.

Particularly important are two commands whose managers made the bulk of the key managerial decisions concerning complexity of the tank: the Army Materiel Command (AMC) and the Army Training and Development Command (TRADOC). The AMC is "the Army's primary agent for acquiring and supporting the [new] equipment" (Keith 1980:2). Established in 1962 to consolidate the Army's technical services, the Army Material Command[4] is given the mission of designing, improving, procuring, and supplying weapons and equipment to the rest of the Army. Comprising 126,000 employees, of which 92 percent are civilians, it has thirteen major subordinate commands, including one testing command, three research laboratories, four schools, and seventeen other activities spread over sixty-five installations and eighteen subinstallations (DARCOM 1981).

TRADOC is the command responsible for ensuring that the doctrine and the soldiers mesh with the equipment and battlefield environment. In 1973 the Continental Army Command (CONARC) was divided into the Forces Command (FORSCOM) and the Training and Doctrine Command (TRADOC). FORSCOM owns all the troops unless they are attached to a unified command; TRADOC is responsible for training the troops and for the doctrine that guides the entire process of preparing for war (Weigley 1984:550, 576). Largely filled with civilians, TRADOC has forty-six major units in addition to its headquarters: five doctrine and analysis centers, one testing center, one experimentation center, five ROTC regions, twenty-five centers and schools, and nine training centers (AUSA 1986:288). most of these subcommands are under the command of general officers, among whom there is much overlap in both doctrinal and training responsibilities.

These numerous commands design the doctrine that paradig-

---

[4]In 1973, the AMC was renamed the U.S. Army Material Development and Readiness Command (DARCOM). The name AMC was reinstated in 1983. This note is included to help any future researchers who may be confused by the change in names.

matically directs procedures, obtains and distributes people, and selects, develops, and distributes weapons and parts. With the approval of the institutional level, TRADOC produces doctrine that indicates certain kinds of weapons are needed. The AMC produces the weapon with the cooperation of TRADOC, which has agreed that the weapon meets both doctrinal and training requirements or restrictions. TRADOC then sets up a program to train soldiers on the equipment. The two commands then hand the weapon and the necessary support to FORSCOM, which gives it to a unit that needs the weapon for its missions. In practice, these actions take place over many months as decisions bounce between the many subcommands of each major command, often emerging simultaneously or in reverse order.[5]

The technical core tends to be the rank and file of the organization, the levels where the product is actually produced. In the Army, it is located within the boundaries of the tactical troops.[6]

[5]The accordance of combat command prerogatives to support commands often promotes bitter struggles in the construction of minor points of doctrine.

[6]Weigley, in his *History of the United States Army* (1984), suggests that the presence of the joint service unified commands has turned the Army into a giant procurement agency. According to his designation, the technical core would therefore be the commodity commands staffed largely by civilians. Although the counterintuitive nature of this notion is appealing, it is not consistent with J. D. Thompson's definition of the technical core as the level at which technical rationality is sensible—that is, where outcome preferences are crystallized and cause/effect relations are well defined (Thompson 1967:149). The outcomes and cause/effect relations of support command activities—e.g., weapon system development and training—are not well defined; how much training or development is enough is problematical. By contrast, the desired battle outcome—winning in a defensive position—is well-known, and previous wars have established a large set of cause/effect relations (FM 100-5, 1976).

Furthermore, although nominally allocated to the unified commands, the battle forces continue to be integral parts of the Army organization. The unified commands are highly fragmented organizations, composed largely of single-service subordinate commands that consider themselves members of their parent service—that is, the Army, Air Force, or Navy. The members of the unified commands rotate in and out of positions in other unified, specified, or Army commands. They wear the parent service uniform and are guided and equipped by their parent services, which heavily influence the detailed operational plans for wartime. The unified nature of the commands is undermined by the traditional biases toward the service having the bulk of the forces in the command in selection of commanders. EUCOM is generally an Army general, Pacific Command (PACOM) is usually an admiral. The relations between the upper organizational levels and the battle forces overseas are disturbed only mildly by the presence of the unified commands. Hence, I place the technical core in the tactical Army in the field, not the commodity commands.

[65]

The technical core consists of five corps and six armies in several theaters spread over eight unified and specified commands (AFSC 1983:2–3).[7] The individuals of the technical core are the integrators who, to produce the desired outcome in wartime, must take the people, the materiel, the weapon systems, the procedures, and the budgets offered.

In the technical core, there are two chains of command. The combat chain of command proceeds upward from the individual to the squad, platoon, company, battalion, brigade/regiment, division, and finally the corps and army. The support chain by which a machine is repaired proceeds from the weapon system to the company or battalion repairers, the direct support battalion at division, the direct support battalion at corps, the general support battalion at corps (the latter two are now called the intermediate repair), and finally to a depot in the theater or in the United States. Figure 5 contrasts the two chains of command, showing the relationship between support units and combat commands. At the lowest levels—battalion and below—the support unit is really a section of the combat unit, but above the combat battalions the support units are separate commands.

The technical core function most likely to first reflect clearly the effects of the increased knowledge burden—the maintenance function—relies heavily on intimate knowledge of the machine. For a military, maintenance is the function that is critical to keeping relatively scarce machines available for battle or returned to battle after damage.[8] A group of German generals from World War II said: "An army's fighting power depends to a large extent on adequate maintenance of its equipment" (Mueller-Hillebrand

[7]A "unified" command is a major command usually responsible for defending U.S. interests in a particular geographical area that contains officers and usually combat forces from all three military services—the Navy, the Army, and the Air Force. In contrast, a "specified" command is a major command responsible for a given function regardless of the geographical boundaries of other commands. For example, Military Airlift Command (MAC) is a specified command that provides air transport for U.S. forces throughout the world. European Command (EUCOM) is a "unified" command responsible solely for battle in Europe. Each military service is required to provide the unified commands with combat forces for use in wartime.

[8]If the Army is limited, both by budgets and in mobility, from buying all the redundancy in tanks or parts that it needs to keep its uncertainty acceptable, then repair of damaged tanks is the next best—and less expensive—alternative.

[66]

Figure 5. Combat and support chains of command

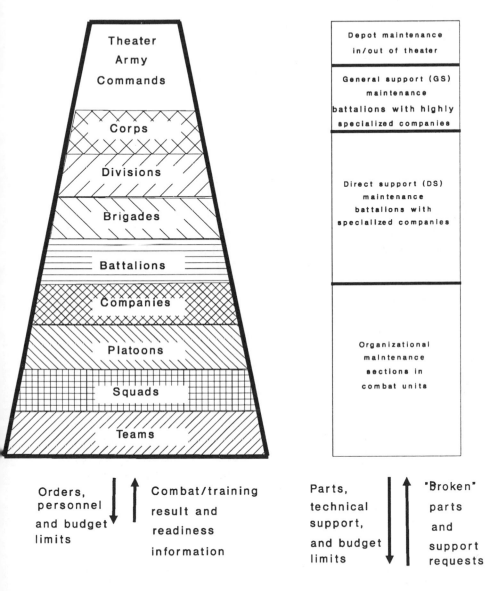

Combat Chain of Command Support Chain of Command

1954:41).[9] Along with the tank, key players on the battlefield, infantry, artillery, helicopters, and so on, represent only the forward tip of an iceberg of support critical to the weapon system or the individual's ability to fight. The rest of the support chain determines the critical outcomes and the organization's ability to respond flexibly to its environment. Figure 6 shows the support chain behind each weapon system in battle.

CRITICAL ORGANIZATIONAL SUPPORT SYSTEMS

Through this chain of support the complexity of the equipment has initial effects that ripple back up the chain to alter the organization itself. An example of such a ripple effect has been described by a German army colonel who discusses what happens when ten infantrymen are traded for ten soldiers—four in a tank and six in a self-propelled howitzer. "When ten infantrymen step into a tank and a gun, all conditions are modified behind them. The Army becomes quite a bit more expensive so that army command will check whether they have to break up units. In addition, the supply requirements of the unit rise sharply. A workshop is needed. Spare parts, ammunition and fuel must be stored in depots, managed and brought to the Front. Finally, the entire supply chain must be led, secured and transported over bodies of water. Where do the soldiers come from . . . [required] for these tasks and who were certainly not at all addressed in the first computation?"[10]

It is also through this chain of support that the efforts of the managerial level to adapt the organization to the equipment in advance of its arrival are felt (see Figure 7). In effect, the ripples due to increased complexity and its increased knowledge burden run both backward from the equipment once it is in the technical

[9]In an entirely different environment—a French firm—Crozier observed that maintenance was central to the organization's certainty and to the resulting relative power of the maintainer (Crozier 1964:109).

[10]See Uhle-Wettler 1981:18. See also Slatkin 1982, whose Congressional Budget Office study argues against higher-quality soldiers based on tank-firing scores, not on the needs of the support chain. For opposite results, see Scribner et al. 1986.

Figure 6. Support chain behind all critical weapon systems in battle

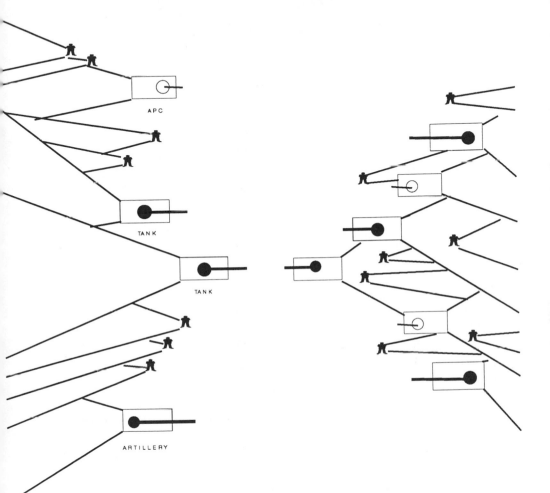

core, and forward in time and sequence from the managerial level commanders and commands.

When there is a shortfall of knowledge about a machine, the first node to feel the effects and respond is maintenance. Maintenance functions are the first alternative to a simple duplication of systems. Being able to repair damaged equipment is usually less

Figure 7. Pivotal role of maintenance in the tactical structure

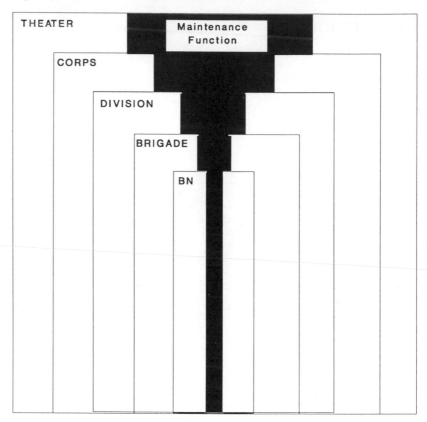

expensive than carrying around a duplicate of each weapon
system. If the system requires unique knowledge in skills or
parts, maintenance units are the first to feel the burden and the
costs. If the system's symptoms are surprising or puzzling to the
humans using or repairing it, maintenance units are the first to
work with symptoms. When there is equipment scarcity, more
maintenance is the way to survive with less in wartime.[11]

[11]Binkin (1986) examines the proportional contributions of maintenance and
inherent reliability to the availability of the weapon system. He offers an elegant
explanation of how improvements in the efficacy of maintenance increase availability more than do commensurate increases in the reliability of the machine.

The chain of maintenance support is structurally layered:[12] the simplest tasks, such as checking the oil and attaching test equipment with Go/NoGo lights, at the unit level,[13] and the hardest tasks, such as major reworking of the weapon system, performed at the depot level. Unit-level maintenance is generally closest to the front lines; the depot level is the farthest away.[14] Between them are two levels: the direct-support (DS) level,[15] whose tasks include more detailed diagnoses than the organizational level and repair using more complicated modules, and the general-support (GS) level.[16] whose tasks are the most complicated in theater and

[12]The structure and levels of maintenance are listed in Army Regulation AR 750-1, entitled "Army Material Maintenance Concepts and Policies." This description of the four basic maintenance levels is taken from the 1978 version, which clearly stated the managerial level's expectations about maintenance and modernization. The level names have changed slightly over the years; the major change has been the nominal coalescing of direct support with general support into a two-part level called "intermediate repair." The units remain differentiated, however, and hence the basic four levels remain.

[13]Unit maintenance includes the repairs allowed to crew/operators and the basic repairers at the company or battalion level. This kind of support is still in the combat chain of command. The work normally includes preventive maintenance, such as oil changes, inspections by sight and touch, cleaning, minor adjustments, diagnosis and isolation of sources of equipment problems, replacement of easily reached bad parts, and evacuation of large items, such as tanks and other combat vehicles (AR 759-1 1978:3-7).

[14]Depot maintenance is the final Army maintenance stop before the equipment is discarded as irreparable. Staffed mostly by civilians, this work is not considered part of the technical core because it services only the supply system and is owned and managed by the managerial level, the commodity commands. Unless located in an active theater of war, depot maintenance is unlikely to be affected by battlefield contingencies or by the other uncertainties affecting the technical core in particular. In general, its missions are major overhaul of equipment ranging from whole weapon systems to modules in support of the Army's inventory, testing removed parts, doing special inspections and modifications of equipment, modernizing equipment, and manufacturing parts provided by the supply system when necessary (AR 750-1 1978:3-9).

[15]The second level of repair, called direct-support maintenance (DS), is performed by specialized support units located behind the combat battalions and attached to the division or corps headquarters. Normally a battalion with three to four companies, the DS-level unit is responsible for the diagnosis and isolation of equipment malfunctions using assigned test equipment, for the repair and basic calibration of the test equipment, for light body repair and repair of broken pieces of equipment that are economically reparable by replacing modules, coils, resistors, and assemblies (AR 750-1 1978:3-7).

[16]The third level of maintenance, the general-support (GS) level, traditionally has the real in-theater specialists located in specialized companies dedicated to types of equipment. The GS level has two broad missions: one to support repairs aggressively far forward at a higher level of detail than subordinate levels, and a

include heavy body work and repair of complicated components.

Of the four, the direct-support level is the critical repair level between the combat units and the rest of the system. The maintainers at this level are ideally veterans of organizational maintenance themselves, as well as graduates of some advanced training. The unit also operates a direct exchange service by which a using unit brings in a component that is broken and directly exchanges it for a good one. If the necessary replacement component is not in stock, the DS unit orders it, and when it arrives the using unit comes to exchange the bad one for the new one (AR 750-1 1978:3-6). Finally, the DS level's mission is to provide "quick reaction" assistance by sending "highly mobile maintenance support teams" to the organizational level to instruct and assist in diagnosis, alignment, and replacement of modules (AR 750-1 1978:3-6). Every combat unit has a DS unit supporting it, and in principle every GS unit has at least one DS unit between it and the combat units benefiting from the GS unit's work.

Finally, it is the division's DS maintenance in particular where the bulk of the organizational changes are most significant. The divisions are the smallest unit the Army's doctrine considers self-sufficient. They are the basic organizational unit for strategic purposes as well as in more detailed war preparations, such as how many division-sized sets of equipment to store overseas, and in doctoral discussions. After General John Pershing unified the combat forces under a division structure in World War I, General Lesley McNair continued in World War II to have the largest standardized unit in the combat forces—the division. As a consequence, the division is enshrined in ways the regiments, brigades,

---

second to provide repaired items and parts for the supply system. GS companies are coalesced in a GS battalion located in the corps rear areas or behind the corps itself. The work involves repair of modules and repair/modification of large equipment for the supply system, heavy body and frame repair, and diagnosis and isolation of malfunctions in equipment down to the internal piece-part level. The repair of modules or black boxes includes grinding, adjusting or aligning of valves and such, and replacing of internal parts that do not require specialized environments. This repair specifically includes repair of printed circuit boards, the workhorse of the new complex equipment. In addition, in a mission emphasized in the later version of the regulation, the GS level was also to provide technical assistance via maintenance support teams to the units in the area (AR 750-1 1978:3-7).

and lower units as well as corps and theaters are not. The division commander fights the battle, the corps commander orchestrates the battle, the theater commander directs the overall effort—but the division is the hub.

Of particular interest, therefore, are changes within the division's maintenance battalion and secondarily in the corps' maintenance and supply elements that support the division. Increases in differentiation here have reverberations throughout the division and the corps; shortages or gaps in knowledge usually result in informal increases in interdependence with the corps level GS unit and/or the contractor or commodity command personnel available. Moving back from the front lines and behind from combat battalions, the DS maintenance battalion's level is the first clear-cut separation of support units from combat units—that is, as maintenance units rather than maintenance sections within a combat unit.

It is also the point at which maintenance and supply separate into two separate units, with the division maintenance unit ordering and receiving its parts through the supply unit. The combat battalions and companies, however, turn in bad parts and order and receive new parts from the maintenance (not supply) unit. The division maintenance units thus serve as governors of maintenance activity both above and within the division, so changes there greatly influence the effectiveness of the entire system.

Given the three organizational levels and the focus on maintenance structure surrounding the complex equipment, it is possible to track the organization's response to the higher complexity of the new tank, especially the effects on the Army's tactical organization. First, there are the changes imposed on the organization by risk-averse managers in the managerial level. These reflect both assumptions about the complexity of the new equipment and then reactions to disappointment in these assumptions. Second, there are the organizational changes in the field among the tactical units, which reflect the shortfalls in knowledge—the rogue set—as the tanks arrive and are put into operation. These changes are demonstrated in the next two chapters, respectively; Chapter 7 then discusses the implications of such changes for operational success.

THE EFFECTS OF COMPLEXITY ON THE ORGANIZATION

If the new tank is more complex than the old tank, the unknowns—the rogue outcomes—of the new tank's operation will impose a larger burden of knowledge on the Army's tactical organization. The costs of this increased complexity would be felt first in the maintenance units of the organization. In these suborganizations, a nonfunctioning tank represents a lack of knowledge needed to keep the weapon system moving. The more difficult the diagnosis and costs of repair, the greater the knowledge involved.

The effects of greater complexity are felt when the organization attempts to meet this increase in knowledge. In efforts to avoid the high costs of complexity, the organization's managers will tend to schedule, concentrate, monitor, and screen the available amount of knowledge. Any remaining shortfalls in knowledge are left for the maintainers to find a way to improvise. What results is a more complex organization. As the next chapter demonstrates, that is what happened when the M1 tank was introduced.

[5]

Assumptions Driving Managerial Responses

For the U.S. Army, the loss in Vietnam redirected the organization to focus on the conventional battlefield in Europe, where Warsaw Pact forces greatly outnumbered NATO forces. For years the answer was to pursue a technological edge over the pact forces, overwhelming quantity with quality. In the late 1970s, senior Army officers directed their organization to seek non-nuclear advanced technologies in pursuit of this edge. In 1977 the army chief of staff stated: "The best deterrence is a credible war-fighting capability. This is measured largely in material terms" (Rogers 1977:23).

Modernization, especially automation, would improve the organization's war-fighting by increasing its speed and accuracy in wartime responses (Hoffman 1976:9, Deane 1976:80). Equipment procurement and research were to have an increased proportion of the Army budget, while operating costs were to be decreased. Battlefield uncertainty would be met with a new army and "maximum use of technical advances." Much later in 1982, a senior leader would endorse replacing soldiers with advanced technologies.[1]

The obstacle was declining budgets. Affording modernization within legislated limits and before congressional support waned[2]

---

[1]See, respectively, Kjellstrom 1976:96, Guthrie 1979:53, and Richardson 1982:252.

[2]For most of its history, aside from times of war, Congress has been loath to allocate funds generously to the military. A ground army suffers most from economizing periods because its components are highly divisible; it is possible to

dominated the other tasks in the list of managers' priorities. With the cost of weapon systems escalating, the overriding task was to find ways to control costs in any budget area that did not affect monies for acquisition.[3] The success of the Army's modernization was defined in terms of the success of individual weapons. The M1 was therefore intended to lead the fleet as a success both operationally and financially. The commanding general of the Army's procurement agency stated that the tank would enormously enhance armor performance, especially speed and maneuverability, while reducing maintenance requirements and increasing availability through better reliability and maintainability of its components (Guthrie 1977:62). It would do more and cost less.[4]

These assertions were well received in the burgeoning support for new electronics and training strategies that promised immediate and long-term savings in two large outlays: operations and maintenance, and personnel.[5] Advanced electronics in particular offered several large benefits: increased speed and reduced variation in operations, greater storage and access capabilities, and less costs over time—all of which constituted a potentially large decrease in uncertainty and a commensurate increase in knowledge for the organization (Deane 1976:80). The managers attempted to save money by increasing reliance on electronics, especially in the critical area of maintenance. The goal was that the M1 and the

---

buy half a battalion, whereas one cannot buy half a nuclear submarine. Hence, it would have been unreasonable for Army planners to believe the generous budgets of 1980 and 1981 would endue long for the Army. In fact, they have not.

[3]Managers in general are less sensitive to the probabilities of outcomes than they are to critical indicators of performance (March and Shapira 1989:86). Hence, congressional interest in costs and production schedules were powerful performance targets for Army managers.

[4]This "overoptimism" was noted by the new commanding general of the Army Material Command in 1982, who was by then beginning to deal with the consequences of this widespread enthusiasm for electronics (Keith 1982:61).

[5]The structure of the Defense Department budget encourages savings taken from personnel or logistics. For every dollar cut from the current budget in personnel, 96 cents is saved; for Operations & Maintenance, Research & Development, and acquisition, respectively, the savings are about 80 cents, 50 cents, and 25 cents. For example, a dollar cut from the program of a multibillion dollar nuclear carrier saves only 4 cents in the current budget. Hence, personnel and O&M are almost always cut the most, and acquisition the least (Demchak 1987: Appendix A; Demchak 1984).

several hundred new systems planned for the Army by the early 1980s would incur no unusual support costs.[6] There would be no changes in logistics organizations and no new management information processing systems (Guthrie 1977:62).

As a major weapon system in battle terms as well as cost terms, the tank started a rippling array of decisions both before and after its introduction. The organization became reacquainted with the problems of scarce knowledge, and in a tyranny of small decisions'its managers initiated a cascading process of complexification in the tactical forces (Kahn 1968). Two broad categories of adaptations produced the overall change in the organization. The first set, called anticipatory changes, occurred before and during introduction of the tank into the inventory. Based on powerful and enduring assumptions about technical complexity, these changes were imposed on the tactical organization by managers determined to get the promised advantages from the newer machines. Later included in this set were the marginal changes the managers made after the problems with the M1's complexity began to be at least partially acknowledged.[7] These all accumulated into a greater complexity in the tactical units.

The second broad set of responses—the reactive changes—occurred after the tank was formally introduced into the tactical units in the field. These reactive changes (discussed in Chapter 6) represent the remaining extent of missing knowledge because they occur after the manager's perceptions and corrections have molded the system around the tank.

---

[6]The actual number of new systems that entered the Army in modernization is difficult to document. In the 1976 *Army Greenbook* the figure is 40 major systems. In the 1979 *Greenbook* the number is 140. By 1980, the most widely published figure was 400, also in the *Greenbook*. By 1982, however, the number the Army chief of staff gave was 583 (Meyer 1984:23).

[7]The components of the system interact with other components in ways that both are and are not desirable for the continuance of the system. At the time a system is initiated, these interactions have still to occur—whether they will result in desirable outcomes is yet to be proven. The system's set of required relations, its "structure," is therefore a hypothesis about the cause-and-effect relations that will emerge (Landau 1979). When managers change the structure, they are changing their collective hypothesis about what causes what when the system operates.

CAUSE AND EFFECT

A vague hypothesis about causes and effects guided the managers' actions: moving knowledge from the expensive, unreliable human being to the more reliable, faster, and seemingly cheaper electronic machine would make operations faster, more accurate, and less expensive, even in wartime.[8] Nowhere was this more evident than in the changes to maintenance in the organization. Here, where the uncertainty of the machine meshes with the uncertainty of the organizations, several presumptions were commonly associated with this move to electronics.

First, it was assumed that the political support necessary to the success of modernization depended on rapid development and production—to the neglect, if necessary, of developing such organizational support as successful training or supply strategies (Kane 1981). Delays meant congressional inquiries, which in turn meant further delays and possible budget repercussions (Demchak 1987: Appendix A). Furthermore, a new doctrine emphasizing quick, accurate, and lethal strikes against a numerically superior enemy was being developed around the promises of the electronic battlefield and being promulgated as justification for much of the acquisition. The doctrine envisioned soldiers fighting highly electronic wars and encouraged the search for complex machine capabilities as solutions (Barnaby 1986, FM 100-5 1976).

The second assumption was that the new equipment would fit into the existing support organizations and procedures with only minor difficulties.[9] A new division structure under development (called DIV 86) was intended to be optimized to the increased capabilities and promise of the new machines. In working out this doctrine, the managers intensified their efforts to balance precise-

---

[8]I have inferred this hypothesis from the sum of the managerial level's actions. It is not likely that members of that level saw their actions as experiments on a grand scale. More likely, they saw their efforts as the only reasonable and rational steps to take under the circumstances. The bulk of these would be seen as minor improvements on the operating system, not experiments in reorientation with large consequences. For further and better reading on this "bounded rationality," see Simon 1976, Williamson 1975, and March 1989.

[9]The neutrality of the new machines for the receiving organization is a common presumption (Posen 1984:55, Steele 1983) See also Van Creveld 1989, Kaldor 1981, and O'Connell 1989.

[78]

ly the number and type of weapons, skills, supplies, and missions in a unit. Such support as maintenance would change, but only to capture the perceived advantages. Training and supply functions were expected to adapt to the new systems as they had in the past. The managers anticipated that a mildly bumpy transition would be eased by their preemptive changes, and effective operations including maintenance would rapidly reemerge—that is, with a highly capable machine things could only get better.

Third, it was assumed that electronic systems posed no more difficult maintenance (especially diagnostic) problems than had been experienced before. A senior officer stated: "Today's tank soldiers must learn to use a laser range finder—they push a button and read a digital number. They are not being asked to build the laser" (Starry 1980:40). In principle, no additional support personnel would be needed because, given some additional tools and rearrangements, predicting or detecting failures in an electronic system would not be more difficult than in an electric-hydraulic system. Furthermore, historical experience with major problems in the past suggested that solutions to design or development problems could be postponed without penalty until the system was under production. A senior officer could say confidently that these systems would be just like any other system and that "the final few imperfections" could be worked out in the field (Guthrie 1979:53).

Finally, there seemed to be no other choice for these managers. In the mid-1970s a decline in recruit quality and quantity meant that less-able soldiers would be handling and repairing expensive machines.[10] Putting barriers—that is, sealed boxes—between these less adept hands and thousands of dollars of advanced equipment would certainly seem only prudent. Sealing the boxes also offered

[10]While influenced later by the loss of the draft, the decline in educability was originally due largely to a lowering of entry thresholds in standard armed forces qualifying tests by McNamara in 1966. By the end of decade, where once 19 percent of whites and 54 percent of blacks could not meet the standards, now only 5 percent and 29 percent, respectively, failed (Weigley 1984:570). Recruiting totals were kept at acceptably high levels, but the quality of the soldier would require considerably more investment in training. Although by 1985 the quality of recruits had risen to the level of the 1960s, the evident decline in aptitude of the newer recruits had come at a crucial period and encouraged an enduring commitment to technological solutions (Binkin 1986:78).

extensive savings in eliminating unnecessary training. A senior officer stated: "True, electronic black boxes are complex, but if the soldier has only to press a button to make them work, they are not complex; they are sophisticated (Starry 1980:40).[11]

Given these presumptions, the generally risk-averse military managers were willing to move knowledge about the workings of electronic machines away from the soldier. They developed the "Smart Machine—Dumb Maintainer" maintenance philosophy with two major prescriptions. (1) that maintenance among deployed units would largely be "remove and replace" operations, thereby eliminating the need for the front-line mechanic to know a great deal about the machine (the maintainer had only to remove the faulty "black box" and replace it with a new one) and (2) that built-in or automatic test equipment (BITE or ATE) would be installed or produced to replace the knowledge the repairer would no longer have.[12]

Because the repairer would not need to know much about the machine, the training program would be shortened by omitting expensive in-residence training on theory or practice. As a substitute, the on-the-job training (OJT) program would be implemented by requiring first-line supervisors to perform and document essential maintenance training in the units deployed in the field. Maintainer courses were reduced in substance, preparing the maintainer only to remove and replace larger components, the "black boxes."[13] If little systems knowledge was required of the

---

[11]I agree with General Starry's understanding of the distinction between sophistication and complexity. An advanced simple electronic system—one with a small knowledge burden—would indeed be sophisticated. Unfortunately, it is also difficult to achieve. In this case the "if" condition did not emerge.

[12]In acting on this overall hypothesis, the Army managers were not alone. The other services also sought new equipment and new ways to save on operating costs. They moved to "remove and replace" maintenance and curtailed education in electronics theory (McNaugher 1989:94, Spinney 1985:32). Similar to the M1/M60 situation, the Air Force's F15 fighter is more complex that its predecessor. The effects were also similar and are included for comparison throughout this and the next chapter.

[13]M1 unit level maintainers at the lowest skill level (10) were given less training than their counterparts on the M60 tank. Given the anticipated capabilities of test equipment, training requirements for the M1 were assumed to be less extensive than those for the M60. The M60A3 (MOS 45N) turret mechanic received 428 hours of instruction, while the M1 (MOS 45E) turret mechanic received only 313 hours—a difference of 37 percent. (Demchak 1983). M1 system repairers at the

"remove and replace" maintainer, the managers did not want to pay for more courses.[14] Also reduced were many schools' entry standards and graduation requirements (Bragg and Demchak 1982.[15] In addition, the cost of training was to be reduced by the judicious use of electronics in the form of maintenance simulators (Marsh 1982:13). Since a simulator that depicted the key cable connections of a machine could be produced, the front-line maintainers would not need to know more about electronics, and there would be less need to have an entire weapon system imprisoned in the training system for practicing diagnostics. Simulators could do more too: they could simulate what a future tank would do. With the use of simulators, it was thought that training could begin *before* the weapon system itself was fully constructed and ready for production.

The reliance on electronics increased in other elements involved with maintenance knowledge. New sets of test equipment, tools, and manuals (TMDE) included a major component of electronic maintenance: automatic test equipment (ATE). Lack of knowledge on the part of the maintainer would be compensated by the information stored in the test equipment. It would precisely identify the source of a failure, thereby avoiding costs in ad-

---

battalion and company levels were given the same weeks of training as repairers for the upgraded M60A3.

[14]The Army training community later reinstated some theory in classes, especially the study of wiring diagrams, as a result of regular complaints from maintainers in the field about a lack of basic knowledge when the test equipment did not provide the correct answer.

[15]By the early 1980s, with 100 the average in intelligence for the broader population, the entry requirements for armor mechanics was lowered from 95–100 to 80–85. For comparison, the following list (from Binkin 1986:15) gives rough estimates of the six aptitude categories used by the Department of Defense:

Armed Forces Category	Qualifying test (AFT) percentile	Equivalent reading grade level
I	93–99	12.7–12.9
II	65–92	10.6–12.6
IIIa	50–64	8.1–10.5
IIIb	31–49	8.1–10.5
IV	10–30	6.6–8.0
V	9 and below	3.4–6.5

Literacy appears to be the most powerful of all the indicators of potential effectiveness in a soldier (Toomepuu 1981:32–33). Evidence is mounting to support the intuitively persuasive statement by Colonel Carl Bernard (USA Ret.) that "the dumb die." See also John English's "On Infantry."

[81]

yanced training, extensive manuals, and inaccurate diagnoses.[16] Built-in test equipment was viewed as particularly desirable because it eliminated external adaptors and the need to transport test equipment.

Similarly with the new data automation, managers could and did seek further economies in more precisely controlled inventories of spare parts. An abundant supply of knowledge in the form of a large inventory of multiple spare parts would at least accommodate the rogue set's knowable outcomes.[17] However, the Army's managers could not count on endless time or money; electronics parts make expensive inventories.[18] In addition, battlefield conditions require speed in repair and, equally important, mobility. Hauling a huge inventory around in order to repair by replacing everything is not a realistic option for a tactical unit, so considerable time was spent calculating the minimally necessary inventory of key parts for each type of combat and support unit and looking for ways to target supplies so little money would be lost in transit time or shelved inventory (Deane 1976:88).[19]

Capping all these decisions was a new division structure intended explicitly to use more effectively the new technological advances acquired with modernization. The introduction of the M1 was meant to be the beginning of a revolutionary moderniza-

[16]Since the 1960s, ATE has been heralded as an alternative to the human maintainer, offering unparalleled speed, accuracy, and convenience in its applications. An Army enthusiast wrote in 1964: "Computerization has introduced . . . a new degree of flexibility, reliability and speed; it has reduced dependence upon human memory and compressed hours and days into seconds. . . . The Army's brand-new computerized 'mechanic,' a footlocker-sized electronic device that . . . does everything except hand the repairman his tools" (Pizer 1967:89).

[17]In principle, an endless inventory would mean that neither the human nor the test equipment need understand the pattern; when all the parts that might be associated with the failure were replaced, the machine should be repaired. This process would, of course, take much time and would have to be iterative, since the act of removal and attachment of parts sometimes introduces failures where there were none before.

[18]In a sample of sixty-three M1 black boxes, the average anticipated cost overall was $7,545; for the unit-level maintainer removing the largest and most expensive items, the average anticipated cost was $16,558. (Cost figures come from a sample of research data used in applying the SESAME spare parts model to M1 parts data in 1981 [SESAME 1979]).

[19]In particular, in response to the skyrocketing costs of electronic repair parts Army managers established a direct delivery supply system for all tactical units. For example, a forward repair unit in Europe would receive its parts *by mail* from a depot or manufacturer in the United States rather than from the repair unit behind it at the next level of command. The previous support chain was a layered set of units, each supporting the units in the front. There is a natural tendency for each

tion of the Army as well as an opportunity to restructure the tactical forces in order to reap the benefits of new sophistication throughout the combat forces. The managers attempted to optimize analytically a low-cost structure of support by precisely allocating skills, test equipment, and parts, weapon system by weapon system. The new structure, called Division 86 (DIV 86), was based on the capabilities and anticipated needs of new armor weapons in particular, especially the M1 tank. In addition, to use the new weaponry fully required a new operational doctrine. As it emerged, the AirLand Battle doctrine was more than just an attempt to promote the newly developing capabilities—it relied directly on the promise of electronics for its own success.[20]

Representing the formal response of Army managers to the rogue set's knowledge burden, this new structure entailed a more complex tactical organization surrounding reduced formal training, lowered aptitude requirements, new test equipment, and carefully controlled parts lists. Developed across multiple managers' offices, summarized in the tables of organization and equip-

---

level to accumulate inventories, since each must service the needs of the subordinate units. The resulting large inventories attract cost-cutters, especially after the experiences with the bottlenecks caused by huge depots in Vietnam. As a contrasting example, the current German army strictly enforces layered support and uses it to push supplies forward to the lowest unit. The oft-cited rationale is the need for any unit to be able to operate alone (Demchak 1988).

[20]In the late 1970s, a support doctrine called Fix Forward was under development as an adjunct to the combat doctrine's new emphasis on speed in operations. It relied heavily on contact teams and good troubleshooting; it was an umbrella concept pushing repair forward. Under the Fix Forward concept, the human skill placed rearward by the managerial level would go forward in a contact team tailored to the specific repair need. The contact team concept addresses the cost-of-parts problem directly. When the subordinate maintenance level was stumped, technical experts from the rear units could use their considerable experience and more precise test equipment to expedite the process.

The ambitious doctrine fell afoul of the field conditions created by the managerial level and the technical core. Traditionally, the maintenance contact team notion was in fact used by missile units, which usually had to visit fixed sites. These units, however, had only one type of equipment to nurture, and it changed relatively rarely. Equally important, missiles and their maintainers did not move much. Precision equipment was not bounced on roads, and missileers were always attached to a missile unit with their type of missile (Demchak 1983). Finally, the commanders of these missile repair units answered directly to the commander of the units served; they were not, as planned in the regulation, one or two levels above the using unit in the combat chain of command. All these conditions are rarely if ever present in the repair of nonmissile or nonaviation systems, especially the generous supply of expensive parts and technical experts. Never clearly defined or implemented, the Fix Forward doctrine is currently alive in name only—as a goal of repairing close to the front lines (Nauta 1983:6-3).

ment (TOEs) and implemented in maintenance supply lists of units, the new structure emerged as an aggregation of unit-by-unit structural changes.

MORE INTERNAL CONTROL OF COSTS AND UNCERTAINTY

Hypothesis 2: The more critical the complex machines to organizational operations, and the more constrained the organization's resources or operational latitude, the more internal control and spontaneous adaptation measures will increase and the more complex the organization will become in response.

The differences in numbers, differentiation, and interdependence from the older division structure reflected embedded presumptions about complex equipment, especially the M1 tank. The changes in skills, TMDE, and parts were concomitant efforts by managers to control the uncertainty by optimizing the balance of maintenance ingredients.

Increased numbers, differentiation, and interdependence among the key organizational elements are indicators of increased complexity in an organization. In this case the elements are maintenance units, people, equipment, and procedures that are critical to the use and care of the weapon system. All three indicators in the Army's tactical organization increased as a result of managerial decisions in response to the perceived complexity of the tank.

Increases in Numbers

The most likely expression of a misunderstanding of complexity's effects is the organization's expectations of numerical requirements. The Army leadership expected to be able to operate a given activity with the new capabilities of the new complex equipment and yet with no increase in people, especially support (Guthrie 1977:58). It could not.[21] By the time the DIV 86 structure

---

[21]Chief of Staff E. C. Meyer commented that he believed he would be saving personnel slots with the new technical capability, but upon taking office he received a "bill" for 40,000 additional slots to accommodate the new communications and automation functions (Meyer 1984:196).

Table 2. Personnel and equipment for old division and DIV 86 structure

	Old Division (M60)	Division 86 (M1)	Change
People			
Overall	19,000	20,000	+5%
Division support			
command	2,500	3,500	+40
Division maintenance			
battalion	1,311	1,562	+19%
Maintainers			
Division tank			
battalions (6)	306	294[b]	−4%[c]
Division-level maintainers			
Three forward			
support companies			
in division	300	255	
Division maintenance			
battalion	127	120	
Division-level			
support total	427	375	−12%
Total (Battalions +			
Divisions)	733	669	−9% division overall
Equipment			
Battalion level			
Tanks	54	58	+7%
Toolkits (track vehicle			
+ turret)	12	17	+42%
Parts inventory	160	215	+34%
Division level			
Repair parts[d]	12,000	20,000	+67%[d]
Corps level			
Maintenance company	22	251	tenfold
Maintenance company	300	551	+50%

Sources: DIV ref.; TOE May 1981; TOE February 1983; Nauta 1983:2-3 and E-27.
[a]Numbers include all 63C, 63N, 63E, 45E positions except recovery vehicle operators and armorers.
[b]Later TOE (1984) had 51 unit maintainers.
[c]Army estimates of required increase in unit-level maintainer slots average 1,700 slots Army-wide.
[d]This increase is due to other new weapon systems as well as the M1. M1-only figures were not available.

fully emerged, the number of people in an armored division had risen from about 19,000 to about 20,000 (Table 2). Most of this increase was in support personnel, whose numbers in the division support command rose from about 2,500 to 3,500 per division (Bigelman 1981:25); the maintenance battalion increased from 1,311 to 1,562.

Ironically, many of the additional personnel were technical inspectors and recovery vehicle operators, not "wrench turners." These new people serve to monitor the progress of repairs, not necessarily to apply advanced skill to the job. In fact, the Army's managers initially planned no increases in personnel to maintain the tank because it was assumed that automatic test equipment would make time-consuming diagnoses infrequent and easy (Kane 1981:4–8). In the DIV 86 tank battalion the number of tanks increased 7 percent (see Table 2), but at the battalion level the number of tank maintainers decreased 4 percent and at the division level it decreased 14 percent (U.S. Army, TOE April 1981, TOE May 1982, DIV Ref. 1979, Nauta 1983:2-3 and E-27).[22] Table 2 shows the changes that made percentage changes grow larger the further to rear the unit is found.

Despite efforts to control costs by minimal inventory, the managers could not avoid the increases in parts that a more complex system requires. Provided by the managers, the parts inventory

[22]The number of tanks rose from 54 to 58, but the number of maintainers dropped initially from 51 to 49 in 1981. After three years in use by units, by 1984 the number increased to 51 again, the same as for the tank's predecessor (TOE February 1983; Nauta 1983:2-3). In the units behind and supporting the tank battalion, the number of actual maintainers decreased from 427 to 375. The drop was from 127 to 120 in the heavy maintenance company of the division maintenance battalion and from 100 to 85 in each of the three brigade support battalions (DIV Ref. 1979; Nauta 1983:2-3 and E-27).

The comparison here is made from generic tables of organization and equipment (TOE) for an armor division circa the late 1970s and the initial DIV 86 TOE for a tank battalion dated 1981 and a later TOE dated February 1983. The specialities counted are 63 E or N and 45 E or N. These figures are difficult to establish firmly because of the adjustments made by local commanders. The formal TOE acted as a basic structure that was changed into a "modified" TOE (MOE) at the request of local commanders. Over time, as a number of commands request similar changes, the base TOE would often be changed to reflect the modified TOEs. The current procedure is moderately different and tied more explicitly to the needs of the new equipment. When scheduled to receive new equipment, the organization's TOE changes only when the new equipment, along with its support material, arrives. Each iteration of TOE, then, is designed directly for the new set of equipment and not by the local commander's evaluation.

for a tank company using the M1 increased 34 percent (Demchak 1983, Rasor 1983).[23] The tank's new test equipment, the STE-M1, with its seven boxes of adaptors and cables, also constituted an increase in numbers, along with such things as the number of vehicle toolkits associated with tracked vehicle repair.[24] The parts requirements of the division and corps support units increased too; parts requirements ballooned in the rear areas.[25] Because these additions required further increases in the amount of personnel allocated to ordering, stockage, issuing, and, importantly, monitoring tasks, supply personnel increased as well.

While there was an increase in the number of people and items associated with the introduction of the M1, the number of individuals assigned to actual repair decreased. This is consistent with managerial assumptions that the necessary input of people in diagnosis was minimal and that saving money by reducing personnel costs was more advantageous than reducing parts inventories to the absolute minimum—that is, relying totally on the direct delivery system.

Increased Differentiation

The introduction of the M1 also increased the differentiation of the tactical organization.[26] The Army managers designated new

[23]The average company moved from 160 items to 215 items. Each unit has a prescribed load list (PLL) and an authorized stockage list (ASL). These lists are given to the unit according to what kinds of equipment the unit has and what mission the unit is going to perform.

[24]DIV 86 changes to the battalion also include a 19 percent increase in other tracked vehicles, a 35 percent increase in radio sets, and a 12 percent increase in trucks, as well as the M1-related increases (DIV Ref., TOE May 1981, TOE February 1983, Demchak 1983, Nauta 1983:2-3 and E-27). These increases compete with the M1-related parts for scarce transportation resources, possibly forcing a trade-off between mobility and adequate stocks.

[25]The division maintenance battalion's repair parts list rose from 12,000 to 20,000 parts, an increase of 66 percent. One corps support maintenance unit in the rear faced a tenfold increase in the number of components it would be repairing on tracked vehicles: from 22 to 221 items. A similar unit's component repair list rose from 300 to 550. While some portion of these increases was due to other weapon systems as well, the bulk were identified by members of these units as being associated with the introduction of the M1 (Demchak 1983).

[26]Differentiation is an indicator of the probability of differentness among elements. A group of 50 people in which 45 are teachers and 5 are doctors is less differentiated than one in which 20 are teachers and 30 are doctors.

[87]

maintenance specialities and additional skill identifiers at the division and corps level. The percentage of senior maintainers also increased. Increases in specialization also occurred via the new test equipment (STE-M1) and the upgrade of the criticality of salvage equipment, such as recovery vehicles. Furthermore, with the DIV 86 structure, both unit and DS/GS repair levels faced increases in the variety of equipment given the numerous new systems under development.

For the M1 in the DIV 86 structure, tank battalion level maintenance specialities were doubled (Kane 1981:4–8, Bragg and Demchak 1982).[27] Now specialized on one system, tank battalion and company maintainers were given military occupational specialities (MOS) that were different from the division and corps level repairers; the former were trained to have shallow knowledge of one system, the latter were trained to have deep knowledge of similar subsystems across weapon systems. Furthermore, whereas previously there had been a dozen or so of these repair specialities, such as 63C or 45R, now there were twenty-eight to be managed and tracked throughout the Army (ALOG 1980:22). Table 3 depicts these changes.

The new tank also brought different training emphases for the division-level and corps-level repairers: more specialization. With the addition of the M1's integrated turret, which held the bulk of the new complexity, turret repairers made up a greater proportion of the junior-level maintainers: from 19 to 33 percent of the junior maintainer force.[28] Given the maintenance philosophy of remove and replace, new maintenance supply specialities integral to a maintenance unit's activities also became more specialized and more senior (ALOG 1984C:38).

Despite the managers' desire to rely primarily on built-in test equipment, the M1 needed additional specialized test equipment and tools (Kane 1981:4-23). Under development for the tank battalion were forty-two other new systems; seven were maintenance related and two were test sets. For the DS level of maintainers there

---

[27]Other complex systems have also experienced this increase in specialization. Aviation repairers are specialized to the particular model of the aircraft series, and the complex M60A2 tank had its own repair speciality in turret repair—45R.

[28]Moving from 8 out of 42 maintainers to 13 out of 40 (Nauta 1983:2-3 and E-27).

[88]

Table 3. Skill differentiation in old division and Division 86 structure

	Old Division			Division 86		
Tank battalion maintenance						
	Repair specialities (MOS)					
	63C	45N	Total	63E/N	45E/N	Total
Skill levels: 30–50	5	4	= 9	11	0	= 11
10–20	34	8	= 42	27	11	= 38
MOS totals			51			49
	Maintenance supply MOS's:					
	76D				76C/P	
30–50	0				1	
10–20	8				7	

Division-level Maintenance:[a]

All DS maintenance support specialities

	Division Maintenance DIV Battalions Maintenance Forward Support Battalion Company (x3)			Brigade Forward DIV Support Company Maintenance (x3) Battalion		
Skill levels: 30–50	24	11	= 35	27	25	= 52
10–20	276	116	= 392	222	101	= 323
Division level totals			427			375

Sources: DIV ref.; TOE May 1981; TOE February 1983; Field 1983–; Nauta 1983:2-3 and E-27.
[a]Includes MOS's 63G, 63H, 41C, 45K, and equivalent MOS for older division structure. Recovery vehicle operators and armorers are excluded.

were thirty new items under development for the brigade support battalion, of which twelve were maintenance-related and six were test sets; for the division maintenance battalion the numbers were, respectively, twenty-nine, thirteen, and six (TOE October 1981, TOE April 1981, Bragg and Demchak 1982:4-15–4-17).[29]

A more subtle increase in differentiation, similar to the increases in skill level among maintainers, was the increase in the official criticality of some equipment associated with the introduc-

[29]The M1 is not solely responsible for the development of each of these systems, but it led the modernization process responsible for the introduction of these new systems.

[89]

tion of the M1. Recovery vehicles, full vehicle toolkits, and wreckers were newly designated as primary mission equipment where previously they had been designated auxiliary equipment.[30] Now units were formally unable to perform their mission adequately unless they could recover and maintain the relatively scarce tank assets.

Increases in Interdependence

The maintenance doctrine developed by the managers encouraged interdependence among organizational elements. In the tank battalion, maintenance sections were removed from the combat companies and replaced by maintenance teams assigned to and controlled by the battalion headquarters.[31] As a result, the tank company now had to rely on the scheduling of the battalion maintenance shop to get key equipment repaired. The new maintenance administration section in the battalion maintenance platoon served as a control node for all maintenance actions (Demchak 1983).

New forward support battalions were established at the brigade level, replacing the forward support companies normally assigned to the division maintenance battalion. In principle, the brigade commander owning a brigade support battalion was to be less dependent on the division for support, including maintenance, but the maintenance support company in this new battalion relied on the division's maintenance battalion for parts and backup support just as the old forward support company did. The new forward support unit, however, now had an additional head-

---

[30]This equipment was moved from the ERC-B to the ERC-A category in the TOE. (ERC stands for Equipment Readiness Code.) The "A" category contains "primary weapons or equipment essential to, and employed directly in, accomplishing assigned operational missions and tasks." The "B" category contains "auxiliary equipment which supplements, or takes the place of, primary equipment, should the primary equipment become inoperative" (U.S. Army TM38-L1711 1984). As a result, this equipment is tracked in unit status reports sent up the chain of command and given more careful administrative attention. That this equipment is actually salvage not fighting equipment indicates the increased importance given repair and salvage operations with the advent of the M1.

[31]Many changes in interdependence stem from managerial-level expectations about the entire population of new weapons rather than just the M1. Only the tank-related changes are discussed here, but there are others.

quarters—the brigade headquarters—inserted between it and the division maintenance stocks and expertise (Bigelman 1981:26).

The addition of this headquarters was important because many of these issues of interdependence are related to supply—in particular, the distribution of authorizations to order, stock, and use certain parts, and the designation of specific supply relationships between units, have powerful effects on the interdependence among units. While repair parts are subject to the same requirements and administrative categories as other supplies, only units with a maintenance mission can directly order repair parts from the supply system. Combat units order their parts through their backup maintenance unit and get the parts later from this unit. In addition providing maintenance expertise, the maintenance unit channels repair parts to the tank battalions (Demchak 1983). With another headquarters, the chain of support became more interdependent.

On a broader scale, as costs of parts rose the Army's managers became more interested in controlling the use of parts through more restrictive ordering or stockage measures.[32] Especially interested in reducing the costs of inventory, managers vigorously promoted the concept of direct deliveries of parts by air from the United States to the combat units. Not only did these deliveries promise to comply with the AirLand Battle doctrine need for speed, but they also could reduce the costs of parts moving slowly through a surface travel pipeline. This procedure, however, greatly increased the dependence of the overseas technical core units on the depots in the United States and the continuity of air operations in wartime (FM 100-16 1983).[33]

[32]The U.S. Army has a "pull" system of supply in which a part is not automatically sent forward, but has to be ordered, and there are limits on the quantity permitted. By a contrast, the German army has a "push" system in which any supply used is automatically reordered for the unit by its supporting unit and automatically sent forward. Furthermore, unlike the American tank unit, the German unit below the division has no budget for parts. It orders what it needs without regard to the total costs. If the part is not in inventory in brigade, division, or corps, the unit waits. If it is in the inventory of any supporting unit, the ordering front-line unit gets the part and the other units automatically reorder for their stocks (Demchak 1988). In the case of the U.S. Army, not only do the local commanders have to watch their quarterly parts budget, they also must wait for an air delivery from the depot.

[33]The debate surrounding the trade-offs between aerial resupply and depots in

In addition, because operating costs of the M1 were 41 percent higher than those of the M60 (Slatkin 1982), there was widespread and increased interest in measures of cost control. For example, the Army, like the Air Force, established a system to track critical M1 parts worldwide. By 1985 the organization was expanding this intensive management to other new systems (ALOG 1985a:40).[34]

⌐ Other more-subtle increases in formal interdependence also occurred. For example, a particular washer on the M1 was lined with a precious metal and costs about $1,200 to replace. Although the washer was easily removed by a unit-level maintainer, the stocks of this item were assigned to the division-level unit for better control of the inventory expense (Demchak 1983). In principle, when this washer had to be replaced the tank and unit maintainer had to wait for the division maintainer to come remove and replace the simple item. The situation is similar for many of the black boxes (LRUs) on the M1. Of a sample of sixty-three LRUs for the M1, the unit-level maintainer was permitted to remove only seventeen of the black boxes (Bragg and Demchak 1982). Induced by cost, interdependence between the battalions and the division increased.

Similarly, the STE-M1 cost about $150,000 a unit (1981 dollars) and was sparingly allocated at six to a battalion, one STE per company maintenance team, and two for the battalion maintenance section. This distribution allowed the battalion to control the reserves of a crucial piece of test equipment for battle damage repair in wartime, making the company even more dependent on the battalion.[35]

---

theater is complicated and often emotional. High inventory and pipeline costs are used as evidence favoring the aerial resupply idea; wartime contingencies and the vulnerability of air lines of communications are used as strong arguments for in-theater depots. Costs tend to be more important in peacetime than in wartime, and hence the former argument tends to prevail over the latter until hostilities begin.

[34] The Army contract for M1 supply and repair parts costs for the first year of production cost $48.8 million for 110 tanks—$444,000 per tank (GAO 1983b:3). The parts contract for the first three years of fielding for the M1 cost $428 million in 1981 (GAO 1981a:64).

[35] Only one of the four tank companies could be resupplied from this reserve before the battalion as a whole had no further buffer in tank test equipment. Independent actions in the companies are likely to be curtailed in wartime, as the battalion seeks to conserve its STE resources, possibly by concentrating the STEs

Increased reliance on maintenance contact teams located in the rear areas also brought about an increase in interdependence among units.[36] The newer maintenance structure requires that all of the division's as well as most of the corps' maintenance units be prepared to provide contact teams to help with repair in forward units (U.S. Army, TRADOC PAM 525-4 1980:43). These teams were intended to compensate for the removal of skills and tasks from front-line maintainers, but this compensation also meant that the front-line units were more dependent on rear maintenance units.

In addition, the corps itself, which was originally intended to be a combat command capable of directing any kind of division in battle, began to become specialized. Corps maintenance units were being tailored to the specialities dedicated to particular pieces of equipment. A corps with M1 tank divisions was slowly losing the ability to support any other kind of division as the flexibility provided by the formerly all-purpose corps eroded (Nauta 1983, Demchak 1983).

Even though the M1 program manager stated in 1979 that the XM1 logistical support concept "is designed to field the tank in 1981 without requiring contractor logistical support" (Kane 1981:4-68), contractors were integral to the introduction of the M1 to units. Because of the high visibility and importance of the M1 when it was fielded, from the beginning of the M1's deployment a contractor's representative was located with each M1 battalion. An AMC representative also in Europe provided a direct line to the depots in the United States (national inventory control points, NICP) for new equipment or scarce parts. At least through 1982,

---

in one place and requiring the companies to bring electronically damaged tanks to this concentration for diagnosis, or electronically damaged tanks could be cannibalized without the use of STE. There is ample evidence of this kind of cannibalization in wartime (Mueller-Hillebrand 1954). Even during the well-resourced 1991 Gulf War, fifty trucks completely disappeared owing to cannabalization ("War at a Glance" 1991:A2).

[36]The contact team concept rests on a managerial-level assumption that division and corps units represent pools of expertise that can be shipped forward in wartime in the form of small mobile teams to diagnose damaged tanks, bringing specialized parts and test equipment to repair the machine quickly. With these teams available, the managers could remove skills from the forward units if they could be provided by contact teams, thereby saving training and inventory costs (Narragon et al. 1984:2-7; "Army Material Maintenance Concepts," 1983:4-4-4-8; Nauta 1983:1-1).

M1 repair-part requests bypassed the intervening maintenance supply and screening procedures and were sent directly to the depot at the theater level (Demchak 1983).

Not included in estimates of expenses, these contractor costs have been high, especially because of the need for facilities and for contractor services (GAO 1981a:47). This reliance on contractor support increased the overall dependence of the tactical units on outside sources of knowledge.[37]

All these changes (see Table 4) increased interdependence overall. The individual units would be less able to operate successfully if their connections to supporting units were disrupted, and it would be more difficult to assess damage from disruptions when so many units were dependent on so many others for critical skills or knowledge. Replacement of or substitution for these would be difficult under any circumstances and extraordinarily so in wartime.

THE RESULT: INCREASED COMPLEXITY

The changes were meant to produce faster, more accurate, cheaper repair among units using the new equipment, especially the M1. Implementation of these changes, however, resulted in a more complex tactical organization.

The managers planned for a smaller number of maintainers and reduced the number of division-level maintainers. By restricting the number of personnel who would be initially faced with the puzzling surprises, and who would have the reduced training, the managers ensured that the missing knowledge was not likely to be available from the average maintainer on the spot at the right time.

---

[37]This technique for acquiring knowledge is neither unique to the Army nor the most employed. The Navy has an average of twenty contractors or civilian technicians out to sea with a carrier (Kaiser and Fabbro 1980:2-17). Navy reliance on civilian contractors for aviation repair increased from 30 percent in 1974 to 59 percent in 1984 (Binkin 1986:122). During this period, Army policy not only permitted but also assumed that contractors would support equipment during initial fielding and longer as needed; virtually all new major pieces of equipment are now supported by contractors for at least the first two years ("Army Material Maintenance Concepts," 1983, Kane 1981:4-69). Other systems, notably the Blackhawk helicopter and the Patriot missile system, are expected to be supported by contractors at the division level and above for an undetermined period (Kaiser and Fabbro 1980:2-8). During the 1991 Gulf War, contractors created special rapid-response task forces to provide support for the Apache helicopters and F15E aircraft during both the build-up and hostilities of the 1991 Gulf War (Bond 1991:59, AvWk 1991:25).

Table 4. Changes in interdependence among units

Tank battalion maintenance
    Increased reliance of tank company on
        battalion headquarters for: maintenance team
                                      parts
                                      maintenance administration
    Increased reliance on division for removal of black boxes
    Increased reliance on contact teams from division for diagnosis and parts
    Increased reliance on contractors
Division-level maintenance
    Increased number of nodes in maintenance support chain
    Additional brigade support battalion headquarters
    Increased reliance on corps contact teams for diagnosis and parts
    Increased reliance on contractors
    Increased reliance on aerial resupply of critical and expensive M1 components
    Increased monitoring by depot-level units of placement and use of critical M1
        components
Corps level maintenance
    Increased requirement to provide contact teams and parts
    Increased reliance on contractors
    Increased reliance on depots for critical repairs[a]
    Increased reliance on aerial resupply for expensive parts

---

[a]Repair of M1 printed circuit boards was initially to be placed in the corps-level units in the overseas theater. However, the repair was so difficult and the equipment so expensive that M1 boards continued to be repaired in the United States. This left the corps-level units more dependent on the U.S. depots for critical board repairs (Narragon et al. 1983).

Someone or something else would have to provide the needed knowledge, especially in diagnosis: perhaps test equipment, perhaps a contact team of senior maintainers, perhaps a large inventory of parts. Later the managers added more senior personnel and more additional training in order to compensate for higher-level skills not placed in the front-line units. These people would not be on the spot for diagnosis but would have to be summoned. A potentially costly delay in maintenance responsiveness was therefore formally built into the system by personnel decisions intended to save costs.[38]

---

[38]One analysis suggests that if the number of maintainers were allowed to rise or fall unconstrained, the M1's repair burden would require an increase in the population of unit-level maintainers across the Army: about 1,700 new spaces required for M1 maintenance across all the M1 battalions in the Army, or about ten to thirty maintainers per battalion (Kane 1981:4-63). If the estimate is correct, these missing personnel needed for the support of the complex machine constitute a shortfall built into the structure of the tactical units to which the tactical organizations would have to respond in order to use the tank reliably.

[95]

In addition, support equipment and parts increased dramatically in their numbers, and the number of support personnel rose as supply requirements increased, especially those individuals associated with the computerized control of inventories. The tactical organization began to attempt to manage closely supplies that had formerly been loosely monitored.

Differentiation increased with introduction of the M1. Training was reduced, making the gap in skills between the front-line maintainers and the division or senior repairers much wider. In addition, the lower level of soldiers' initial abilities and formal training also reduced the future capability of the soldier to deal with a large rogue set.[39] The ability to reconstruct a probable pattern of events from fragmentary data is crucial to accurate diagnosis of a maintenance failure, since electronics systems usually exhibit few hints about the failure save an unwillingness to operate properly. Linking a lower mental ability in front-line maintainers with integrated electronics systems meant that the knowledge needed to accommodate the vicissitudes of the rogue set would not be located in the maintainers but would have to be located elsewhere.

Other changes were not planned so much as unavoidable as the training elements of the managers became acquainted with the new complex machine. Despite the high expectations about automatic test equipment (ATE), the new tank still required new specialities in both repair and operation, plus extensive additional training of division repairers and their supervisors (Demchak 1983).[40] Equipment also increased in both variety and number with

---

[39]It is not yet clear how human beings think. We do know a bit about how college students retain and gain access to their memories. They receive and create patterns of information much like the interconnected branches and twigs of large trees. Following a chain of reasoning is much like following a tree from its trunk out a branch to a far twig (Fitts and Posner 1967, Margolis 1987). Individuals with less capability appear to have greater difficulty absorbing and, more important, inferring patterns.

[40]In the original DIV 86 structure, the Army's managers reduced the number of junior maintainers from 42 to 40 at skill levels 10-20 and provided no compensating change in senior repairers. After two years of experience with the M1, however, the M1 tank battalion received more senior maintainers, moving from 9 to 11, a 22 percent increase (DIV Ref., TOE May 1981, TOE February 1983). In the old division structure, senior maintainers constituted 8 percent of the population, but in the new structure they are 14 percent—an increase that occurs mainly in the

the introduction of the M1, producing a greater need for supply specialities and additional training on the new logistics requirements. In the late 1980s, the test equipment acquired its own maintenance speciality—a test equipment "tester."

All these increases in differentiation imply a loss of overlap among the elements of organization. When the maintenance specialities multiply, the likelihood of a maintainer understanding a random unit's piece of critical equipment decreases across the population, and hence more organization resources have to be spent to allocate more precisely these more tailored maintenance skills, in wartime as well as peacetime. The problem is not that the organization cannot find a way to do this allocation, but that the costs of establishing and continuing a highly precise personnel distribution system are high and not likely to have been included in the original cost estimates for the weapon. The costs of not precisely allocating the personnel can also be high; they are found in the number of tanks down and good parts mistakenly sent into the repair chain as bad. Either way, the costs must be borne.

Increases in interdependence occurred in a variety of often seemingly unrelated changes. Scarce people and items were concentrated at the battalion headquarters, making the line companies much more dependent on battalion coordination activities and priorities. The supply units behind the M1 tank battalions became more interdependent in that an additional headquarters, the brigade support battalion, is now involved in the maintenance decisions of the division's units; the brigade support battalion's priorities will direct the interactions between its forward support maintenance companies and the division maintenance battalion.

Managers' efforts to substitute for scarce skills have led to increased formal reliance on automatic test equipment, contact teams, and contractors. As human troubleshooting skills were significantly reduced in the training for M1 battalion and company maintainers—the first to see a damaged tank[41]—reliance on

---

heavy maintenance company of the division maintenance battalion (DIV Ref., Nauta 1983:2-3 and E-27).

[41]In the basic course for maintainers, the number of tasks assigned to troubleshooting declined from six for the M60A3's 63N and five for the 45N, to four for the M1's 63E and two for the 45E (Bragg and Demchak 1982:4-7).

[97]

automatic test equipment was assured. Unfortunately, however, test equipment cannot have a leap of insight when faced with an unforeseen outcome, and the hasty addition of the test equipment did not include a redesign of the tank itself to increase the number of testing access points available. Because the test equipment was unlikely to diagnose the difficult failures adequately, the harried maintainer was left to wait for help from rear units or to begin replacing everything until the tank worked again. The lack of redundancy and depth in the skills at the front line encouraged increased dependence either on contact teams or on the supply system.

The difficulty with tailored contact team concepts, of course, is the extent to which these teams have this expertise, mobility, parts, or personnel to perform contact team services. The reduction in training and in the number of division maintainers made the needed expertise less widely available and therefore a scarce asset. The costs of parts alone made it less likely that a support unit commander would be willing to send a contact team truck carrying expensive parts away from the division unit to the uncertainties of the front lines. In peacetime, contractors can provide some of the missing diagnostic expertise and parts, but in wartime this expertise and the parts are not likely to be available. Contractors encourage a more subtle habituation to a source of expertise that disappears just when it is most likely to be needed—in wartime. Contact teams, the tailoring of corps units, and contractors have made all the units involved in the maintenance of the M1 more interdependent for key skills, diagnostic equipment, and parts.

Interdependence also rose with the increased criticality of some equipment; test equipment in particular became an indispensable part of maintenance rather than an aid to accuracy. Given the decisions on training and the complex design of the tank, the viability of all the other money-saving mechanisms would ultimately depend on the performance of this equipment and the ability of the supply system to respond to unforeseen demands for expensive parts.

Organizational promoters presented ATE as a virtual panacea for all problems of weapon system repair that stemmed from the complexity of the weapon.[42] Although dampened to some extent by experience, the enthusiasm for ATE nonetheless survived through the 1970s. Persuaded that ATE and certain kinds of training could compensate for most equipment design deficiencies,[43] the managers tried, albeit unwittingly, to avoid paying the price for the scarce knowledge designed into the new complex machines.

The problem was that this test equipment had to be programmed to find failures, so in reality it could only find known failures. ATE can identify the surprising failures only indirectly— by identifying known fragments of the overall pattern of events. More of these fragments could have been known, but that knowledge required a willingness to invest early in expensive research on the part of the Army's managers. Furthermore, this test equipment is of use only after the failure has occurred; it cannot predict failures, no matter how many times it has identified pattern fragments.

Without the human capability to construct patterns creatively, the test equipment does not provide the right kind of missing knowledge and is also complex and expensive. Electronic test equipment cannot identify faults that emerge in the connections between electronic boards, the critical "interfaces" that link the chip to the mechanical components. The rogue set is affected only mildly by the test equipment, and without human sources of knowledge the remaining possible maintenance alternative is an abundant and expensive supply of spare parts.

In addition to test equipment, recovery vehicles for salvaging damaged tanks became a part of the minimum essential equipment for wartime; tanks were too precious simply to push to the side of the road and abandon. Toolkits, now specialized for a particular tank system, were more expensive and less interchangeable; they were also elevated to the list of critical equipment. In short, much more material had no handy substitute and

[42]For example, when the M1 tank was discovered to have unexpectedly high internal complexity, the chosen answer was not to rework the design; instead, ATE was hastily ordered to be developed for the M1 (Bragg and Demchak 1982).

[43]In the face of criticism of the M1 tank, Army representatives always insisted that later improvements, called Product Improvement Packages (PIPs), will take care of any remaining shortcomings of the basic M1 design (Binkin 1986:90).

[99]

had to be more carefully managed in order to have that capability in wartime.

The same criteria applied to expensive spare parts. Managerial efforts to reduce overseas inventory costs of scarce parts and to direct the deliveries of parts more precisely resulted in worldwide monitoring of M1 parts in units as well as an increased reliance on aerial resupply and U.S. depots. This is particularly pernicious when the front-line units are unable to diagnose failures adequately. Puzzled people pull parts, and often they pull good parts, throw them into the repair bin, or introduce new sources of failure by the act of pulling the parts. During at least the first few years of the M1 in the field, scarcities in M1 parts were widespread, in part because maintainers were fixing failures simply by pulling anything that could be remotely related to the symptoms displayed.[44]

The natural tendency to repair by replacing everything increased the already increased costs of the M1 and its support and encouraged close monitoring of parts usage and allocation.[45] The Army now tracks the location and allocation of expensive parts for several major systems, in addition to the M1. As the intent is to control the flow and the costs of parts, tactical divisional operations require more investment in coordination than they did before the M1 was introduced. And many more relations have to be functioning well for the operations to be successful.[46]

---

[44]This is a common phenomenon in electronics repair where the system is difficult to diagnose. A good indicator of this lack of knowledge is a figure usually collected only in the rearward support units and not collated elsewhere: the No-Evidence-of-Failure (NEOF) percentage. NEOF is the proportion of putatively broken components that arrive at a rear area support shop for repair but show no evidence of being broken. A high NEOF percentage means the front-line units are simply replacing boxes in order to fix puzzling failures. One corps unit dealing with electronics had a 90 percent NEOF for three months (Demchak 1983).

[45]The Air Force, from its inception a highly technical service, is not immune from puzzling failures. In 1981, after several years with the F15/16, Air Force maintenance units recorded an Unnecessary Removal (UR) rate of 40 percent, the equivalent of the NEOF rate (Rue 1983). Nearly half the time a maintainer removed a box and sent it in for repair, there was nothing wrong with it. Given that this occurs in an explicitly technical service, there may be a fundamental limit to diagnostic accuracy closer to 30 percent than the 10 percent or less included in military specifications.

[46]The Air Force has had similar experiences with the F15. Its numbers have increased: from 1975 to 1983, maintenance personnel increased 45 percent for the

THE LIKELY RESPONSE IN THE UNITS

The sum of the managerial decisions aimed at controlling costs increased the formal *N, D,* and *I* of the units affected by the introduction of the M1. Facing unknowns, the managers reacted much as expected: they concentrated, scheduled, restricted, and monitored. Once in the field, the M1 would then pose something of an enigma when the knowable or unknowable unknown outcomes occurred. Front-line maintainers would not be prepared to diagnose those outcomes, given the complexity of the tank, their shallow training, and limited inventories of parts. In addition, the scarce and limited test equipment would increase the knowledge burden by its own internal complexity.

The front-line soldiers, even less able than the managers to tolerate increased uncertainty, would also react by concentrating, scheduling, restricting, and monitoring more intensively. These responsive changes then produced a more intricate tactical organization with a yet larger rogue set itself. The next chapter demonstrates the second half of this process of complexification.

APPENDIX TO CHAPTER 5:
CHANGES IN DIFFERENTIATION IN OTHER SYSTEMS

The M1 is not unique in these anticipatory changes; other more complex systems exhibit the same influences on the planning and

---

F15 (Kaiser and White 1983:G-15). The F15 has four times the number of internal components of the earlier and less complex F4, and its essential test computer has 130,000 parts. In terms of differentiation and interdependence, maintenance depends heavily on scarce and unique computer assets. For the 45 black boxes per aircraft that intermediate repair addresses, a wing has 3,240 boxes but only six computers. The average diagnosis takes three hours, and in 1979, the test computers themselves were operational only 50 percent of the time, improving to 80 in 1980. The average time to fix off-base failure in the F15 was nine times that of the F4 (Spinney 1985:32-36).

The reliability problems have prompted the tactical air wings to devote considerable efforts to managing intensively the repair of F15 components, monitoring test equipment availability, using known good boxes to supplement diagnosis, and relying on contractors (Kaiser and White 1983:G-12). For 1980, manufacturer representatives logged man-year totals that were 53 percent greater than the nearest competitor, the F/RF-4 (Kaiser and Fabbro 1980:2-10). Nonetheless, nearly a quarter of the black boxes replaced were good boxes. The aircraft's low reliability plus hemorrhage of good boxes mistakenly replaced caused a spare-parts shortage in 1980-82. The backlog was not fully met until 1984 (Kaiser and White 1983:G-6–G-7).

subsequent structures at the managerial level. A number of major weapon systems show complexity changes similar to those of the M1. In a sample of twelve systems,[47] the length of operator training increased for half, held constant for one-third, and decreased for two. The length of organizational-level training had increased for five, held constant for four, and decreased for three. DS/GS training had increased over eight systems, held constant for two, and decreased for two. Given the predilection to shorten training, it is likely that the systems for which training increased were systems in which there was no reasonable or other way to avoid the increase. The bulk of the systems experienced an increase in training time—that is, increases in the knowledge burden.

There was also an increase in the number of specialities specific to weapon systems. Operators in six of the systems have their own specialities, while operators in two other systems are required to have an additional identifier before they can operate the equipment. Of the nine systems that require organizational maintenance, the five have organizational-level system-specific specialities. Expected to repair a range of equipment, the DS/GS repairers do not have system-specific specialities. For the M1, the lengthened courses previously mentioned indicate the increase in knowledge requirements; similar figures undoubtedly exist for most of these other systems.

While the data was not available to compare all the other *N*, the *D*, or *I* changes across systems, this and other fragmentary evidence suggest that the knowledge requirements have increased for these other systems as they have for the M1. The consequences for the receiving units are likely to be similar across units.

---

[47]These include the Bradley Infantry Fighting Vehicle, the Roland air defense missile system, the improved TOW Vehicle, the UH-60 Blackhawk transport helicopter, the Stinger portable shoulder-fired cable-guided missile system, and the Patriot medium and high altitude missile system.

[6]

User Responses to Complexity

With complex equipment and a large knowledge burden, a constrained organization sprouts a myriad of spontaneous "quick fixes." These reactive changes are responses to the unknowns or missing knowledge remaining after the manager's perceptions and corrections have molded the system around the tank. Properly viewed, they contain feedback from the tactical units about the tank's complexity. Members of units associated with M1 tanks tried informally to fix their knowledge problems, but in doing so made the organization still more complex. Both the unit members and the managers were dedicated to "making it work," and both produced more fragile, interdependent relationships.

Military units in particular are encouraged to survive no matter what the circumstances. They tend to innovate around problems locally in ways that are often not formally blessed. Like the risk-averse managers, the members of a deployed unit seek to accomplish the mission in any way possible, despite the knowledge burden of the tank. These adaptations tend to grow into informal norms and procedures of operation that become crucial to success.

A "can do" attitude is a highly desirable organizational ethic in uncertain operations such as war, but it does have a disadvantage. Rarely discussed and even more rarely seen in print is the fact that the local adaptations can change the true capabilities of the force. A multitude of minor variations autonomously appear in the tactical forces as each individual unit and section makes arrangements to accomplish its own missions. There emerges an interconnected web of relationships and dependencies that work as long the coordination and resource interactions are not significantly disturbed. The result is a seemingly robust peacetime organization whose ability to survive disruptions of these relationships in wartime is open to question.

KNOWLEDGE SHORTFALLS IN THE FIELD

Hypothesis 2: The more critical the complex machines to organizational operations, and the more constrained the organization's resources or operational latitude, the more internal control and spontaneous adaptation measures will increase and the more complex the organization will become in response.

When the new machines reached the tactical army, key aspects of the "Smart Machine, Dumb Maintainer" philosophy did not meet the expectations of the Army's managers. The scarcity of knowledge produced a multitude of imbalances and bottlenecks for the technical core. In the areas of TMDE (especially automatic test equipment), maintainer skills, and structural assumptions about required parts supplies, the managers' initial calculations fell seriously short of the field's requirements.[1] In short, the units did not know enough, and given the complexity of the equipment, they would not be able to adapt around the shortfalls.[2]

Test Equipment and Training Substitutes

Automatic test equipment (ATE) performed poorly—for example, the use of "simplified test equipment" (STE) was required for diagnosis of the M1. But diagnosis with the STE was time-

[1]Results for other systems, especially helicopters and missiles, are not shown because those results were similar, but examples from these other systems are used in this discussion.
[2]Data presented in this chapter were gathered in part through interviews in 1983 with representative units at each level of the chain of support, starting with both M60A3 and M1 tank battalions and proceeding through divisional, corps, and CONUS support units or elements. These are cited as Demchak 1983. See also Narragon et al. 1984.

consuming, taking two to four hours, and this contributed directly to the not-mission-capable (NMC) rates experienced in the field
(Demchak 1983). The test results were ambiguous, and the inaccuracy increased when the engine was hot, a typical battlefield
condition.

The STE had difficulty diagnosing the tank, especially the
multiple source failures endemic in complex electronics. Generally, STE could identify the sources of failures only one at a time
and serially. Once STE identified a component as bad, the component had to be replaced before any more troubleshooting could
continue; if the part was not available, troubleshooting stopped
until the part could be acquired. The maintainer's typical response was to avoid using the test equipment.[3]

The inaccuracies of the STE also affected the support system,
increasing the dependence of the lower-level units on the parts
supply system. In 1982 a tank was driven into the motorpool of
an M1 unit with a symptom: low engine idle. The mechanics
consulted the STE manuals and selected a test that required thirty
minutes to hook up and run. The test was run five times. After
the first four runs, a new "bad" box would be identified by STE
and replaced by the mechanics. Each time, however, the symptom would persist. After the fifth run, STE pronounced the tank
fixed, but the symptom was still present. Twenty-four hours after
the tank arrived in the motorpool, the frustrated mechanics pulled
the engine and transmission. Finding that a small ball on a pin
holding a part of the fuel management system was malfunctioning,
they replaced it. The symptom disappeared (Demchak 1983). Four
possibly good boxes went into the repair system, and twenty-four
hours were lost; the engine and transmission could have been
pulled without STE.

This frustrating experience was also common with built-in test
equipment (BITE). For the Air Force's F16A, the BITE exhibited a

[3]See Sarna 1982:17. For most of its operational life, the STE has been able to
diagnose only single-source failures. Once STE has identified a component as bad,
the component must be replaced before any more troubleshooting can continue; if
the part is not available, troubleshooting stops until the part can be acquired
(Demchak 1983, Bragg and Demchak 1982). Ten years after STE development, the
extent to which it successfully diagnoses multiple source failures is not clear.
Indirect evidence suggests that it is not a reliable piece of equipment (Tice 1988).

45 percent false-alarm rate; the radar in the Navy's F18 radar has a 45 to 62 percent false-alarm rate. The BITE in the Army's Multiple Launch Rocket system (MLRS) correctly identified faults only 15 percent of the time, far from the 90 percent goal planned by the Army's managers.[4] The M1's BITE performed so poorly during initial tests that external test equipment, the STE, was urgently developed as antidote.

ATE was often difficult to use and prone to failure. A typical task with the STE, its seven boxes of cables, and numerous manuals occupied seventeen pages, while manual alternate test procedures involving the use of a digital multimeter, wiring diagrams, and human skill cover the same task in four pages (Demchak 1983).[5] Furthermore, the test equipment is often as complex as the equipment it is testing. The intermediate test stations used on F15 aircraft have 220,000 parts—two times the parts on the F15 itself. As with any system, this complexity was a net addition to the knowledge burden presented by the new machine.[6]

Training substitutes did not produce adequate maintainer skills for repair of electronic systems. When these systems fail, they give maintainers no external clues about the probable source of the failure. Unlike hydraulic-electrical systems, electronic systems do not gracefully degrade, emitting sounds or giving visual evidence to inform operators or maintainers of the nature of the problem. Steps in using test equipment often involve intricate series of decisions and the use of large manuals. When during operational tests the M1, a broadly integrated electronic system,

[4]In 1980 the F15 intermediate test stations failed every thirty-four hours (Binkin 1986:60-62, 65).

[5]It is difficult to design a machine that minimizes the Type II error (no false positives) without risking increases in false negatives (Type I errors) (Binkin 1986:59). The high rate of false alarms, however, is an example of the ways in which ATE increases rather than decreases the universe of outcomes in the system. Without the ATE, the squadron would not have spent the afternoon hunting for the failure in the grounded aircraft. Such high rates bode ill for wartime operations when parts are likely to be scarce.

[6]See Binkin 1986:60. The General Accounting Office and others have argued that a separate speciality purely for the testing of test sets, especially the STE-M1, is needed (GAO 1981a:53). By the late 1980s, the Army had announced plans to investigate using test set testers.

stopped functioning, two equally equipped maintainers blamed opposite ends of the tank.[7]

If the ATE does not identify the origin of the problem, the maintainer must use a mental library of possible patterns to make educated guesses. Reductions in training, however, especially in theory, leave maintainers without the necessary skills or experience.[8] The alternative was serially replacing all components suspected of contributing to a failure.

The use of ATE did not compensate for moving expensive knowledge to the rear in order to concentrate or monitor its use: squeezing a balloon at one end makes the other end larger; it does not eliminate the air inside. As long as the organization has high requirements for speed and accuracy in its responses, it cannot avoid paying the scarcity of costs of the knowledge burden by using ATE.[9] Similarly, moving the knowledge burden rearward using ATE and less training does not eliminate the uncertainty requirements. As shown in Figure 8, that knowledge must still be paid for in the system.

Maintainer Skills

On-the-job training (OJT) also has been not as successful as hoped. It was not anticipated that skills would be less transferable than expected between the old and the new equipment. For example, one key source of training savings anticipated by the Army's managers was a direct equivalence in tasks between the M60 to the M1. These common tasks would therefore not need to be taught in M1 courses given to M60 mechanics, and money would be saved with shorter in-residence courses.

These tasks also represent the level of commonality that managers assumed existed between the two tanks. When individuals were tested on the M60, trained on the M1, and then retested in

---

[7]See, respectively, Sarna 1982, Wohl 1980, and Bragg and Demchak 1982.

[8]One survey showed that graduates of basic Navy electronics repair courses did not know what to do next if the ATE did not locate the correct source of a failure (Binkin 1986:60). This result is consistent with the observations by Sarna (1982) and the theoretical findings of Wohl (1980).

[9]STE-M1 costs $150,000 each in 1981 dollars (GAO 1981a:53).

Figure 8. ATE's incomplete displacement of the knowledge burden

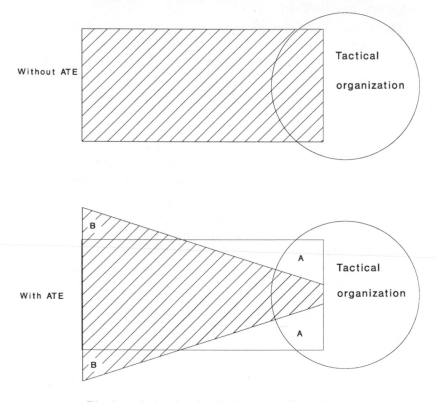

The knowledge burden in A is not eliminated by
moving to location B in the support system.
The overall organization must still acquire this knowledge.

September 1981, the equivalence of tasks appeared dramatically lower than anticipated. For the turret mechanics (MOS 45E or 45N) normally located at the company or battalion level, only one-quarter of the tested tasks transferred as expected. For the tank operator—the job that seemed to require the least technical skill—only 20 percent of the transfer tasks transferred acceptably; some of the tasks had a 10 percent pass rate (Maitland et al. 1981:4-4).

Because of this low level of commonality and compensatory training, supervisors lacked the necessary expertise to supervise

[108]

or teach the maintenance novices. In the early 1980s, managers began sending training teams out to the units to "train the trainers" because of the low quality of OJT reported back from the field.[10]

Gas-turbine engines like that of the M1 require more technical ability than the managerial level foresaw. In addition to receiving additional skill identifiers for the M1, tank subsystem maintainers at the direct-support level were required to attend special training. The basic normal training of direct-support maintainers included all tactical tracked vehicles in the division. The additional training for the M1 alone, however, was one-half to two-thirds as long as the training for all the other vehicles. For 45K, the change was an additional six weeks over the basic ten weeks in 1982; for 63H it was six weeks over the basic ten.[11]

Simulators also have been less successful than desired. One major objection was their costs: a driver trainer with five trainer stations cost $4.1 million in 1981—the cost of nearly two tanks at accepted prices of $2.7 million each, or of four tanks at the Army's official per tank cost inflated from the $500,000 price in 1976 dollars to $750,000 in 1981 dollars.

The chief complaint, however, was the lack of realism or "fidelity."[12] For example, tests on the Bradley personnel carrier indicated that those trained in "live" fire exercises were better at hitting targets when the machine was moving than those trained on a simulator—89 percent to 61 percent, respectively. Simulator-trained gunners see targets slightly more quickly but hit them more poorly, suggesting that the difference lies in adjustment to

[10]Ironically, students themselves, all of whom were experienced M60 mechanics, have requested that the more difficult tasks be taught in the resident courses and the easier tasks be left to on-the-job training (Nielsen 1983). This request conflicts with the cost-saving impetus behind the OJT program, the use of ATE, and the "remove and replace" training philosophy (Maitland 1984:5-1).

[11]See, respectively, Binkin 1986:9 and Bragg and Demchak 1982, chart 3-5. Increased requirements for training have emerged on other systems. A mechanic for the Patriot air defense system now must learn 605 tasks in 38 weeks while the previous system, the Hercules, required only 19 weeks and 112 tasks (Binkin 1986:44).

[12]See, respectively, GAO 1981a:74, Maitland et al. 1981, and Bragg and Demchak 1982. Col. Bill Donovan, a retired Air Force decorated combat pilot, observed that he felt he had to relearn how to fly after spending four weeks on a simulator (Private conversation, Summer 1984).

the feel of the machine versus the smooth operation of the trainer. The attitudes of the soldiers tested also indicate a difference. Although they generally approved of the simulator, the noted simulator deficiencies were the unrealistic representation of recoil, noise, and obscuration of the target. Finally, an objective deficiency of the simulator was its tendency to present targets too distinctly, quite unlike battlefield conditions.[13]

Structural Assumptions

Several of the managers' assumptions combined synergistically into larger problems. For example, the low aptitude levels of the maintainers exacerbated shortcomings in test equipment and training substitutes when the machines reached the tactical forces. Although the overall Defense Department distribution of aptitudes returned to the levels of the draft years, recovery did not occur in the maintenance specialities.[14] In 1968, 60 percent of the personnel in tracked vehicle repair scored above 110; the minimum entry level was 90 (Brachman 1968:III–40). By 1982, 65 percent of the personnel assigned to tracked vehicle specialities were in the lowest mental categories (Demchak 1983). The standardized entry exam (AFQT) scores for M1 maintainers have remained lower than those of tank crewmen or most of the other soldiers (Binkin 1986:64). In a system where a score of 100 was roughly average, personnel in key direct-support specialities tended to score 15 points below the average—that, is 85 (Bragg and Demchak 1982).

Furthermore, performance standards were unrealistic. The optimistic structural assumptions about performance that managers used in designing support relationships and supplies were faulty. Estimates of maintenance manhours, diagnostic and repair times, and reliability were low. In tests conducted in 1981, maintainers required more time to diagnose with the STE-M1 than would be allocated according to the M1's maintenance allocation chart (MAC),

[13]See Butler 1982:4-2, 3-3, 3-15, 3-16.
[14]The changes from the draft era to 1985 are as follows (Binkin 1986:15–16):

Category	1960–1973	1974–(1977–80)–1985	Reading level
I & II	38–35%	35–(29)–38%	grades 10–12
III	49–45%	55–(43)–51%	grades 8–10
IV	14–22%	10–(28)–11%	grades 6–8

one hundred minutes on average versus the anticipated sixty minutes. If accepted as Army standard and used to alter the MAC, this 66 percent increase in required diagnostic time would have far-reaching effects on the allocation of tasks among the organization's levels (Maitland 1981:5-1). Similarly, if the initial estimates were seriously off, then units are likely to be severely imbalanced. Units are likely to have a good deal of difficulty accomplishing the required tasks with the assigned set of skills, people, and equipment.

The managers' estimates of times for tasks and associated requirements for supplies were unrealistically low. The assumptions of diagnostic time required for the use of ATE, six to twenty-five *minutes* per component ("box"), were quite optimistic. A military service with a long history of experience with complex technical machines, the Air Force as recently as 1981 reported two to six *hours* for F15/16 boxes, and the Navy averaged one to three *hours*. The Air Force then averaged between 25 percent and 40 percent mistaken removals of components for its F15/16 fleet (Spinney 1985:32–36). In 1983 the F15's fire-control system had the lowest reliability of any subsystem on the aircraft: nearly 30 percent of the boxes pulled off the machine were good boxes. During the same period, the cannibalization rate *in peacetime* averaged 40 percent over the period 1980–82 (Kaiser and White 1983:G6–G7). In addition, diagnostic accuracy of new ATE was assumed by the Army's managers to be 90 to 95 percent. The Air Force also initially assumed 90 percent for its F15/16 ATE and has experienced over time an average of 60 percent (Spinney 1985:52).

For a number of systems the managers of the Army and other services understated another key estimate: manpower requirements. As a proportion of military equipment, electronics rose from about 10 to 20 percent in the 1950s to about 40 percent in the early 1980s. Similarly, the proportion of electronics specialities across the Defense Department rose from 6 percent in 1945 to 21 percent in 1985. The proportion of technical workers also rose from 13 percent to 29 percent over the same period. For the Army, the figures were from 10 percent to 19 percent, and 10 percent to more than 25 percent, respectively. Despite the trend in growth of technical skills and the massive modernization, however, the

[111]

Army managerial level estimated only minor increases in highly or moderately technical skills over the period 1984–90: 3.7 percent and 2.4 percent, respectively (Binkin 1986:6–8, 38).

This tendency to underestimate was apparent in the manpower projections of the Blackhawk helicopter. Labeling the Blackhawk development as a model program in its attention to reliability, managers planned the helicopter's support with the lowest initial estimate of necessary maintainers: six instead of nineteen in a platoon. Now in the field, the helicopter requires twenty-four individuals to maintain the machine, an estimate error of a factor of four (Moore 1986:31). In November 1990, an internal Army review of maintenance of the Apache attack helicopter concluded that the maintenance personnel assigned to each helicopter battalion had to be doubled to achieve acceptable readiness rates (Baker 1991:27).

The Army's managers were not alone in planning on unobtainable goals. In the Navy, the Spruance class destroyer DD963 was intended to operate with 224 enlisted sailors, but after five years in operation the ship needs 295, a 32 percent increase (Binkin 1986:41). Whether the estimate error was fourfold or 32 percent, the effects across the service were a severe shortage in necessary skills, making it necessary to rely on outside sources of expertise.

One symptom of these inaccurate estimates in manpower was a dramatic increase in maintenance manhours that the managers clearly did not expect. In the Air Force, the F111D aircraft requires five times the maintenance manhours of the much simpler A-10. In the Navy, the F14 requires 3.5 times the manhours of the A-4M aircraft. For the F15 aircraft, despite the decrease to three flight-line specialities from the twelve of its predecessor, no maintenance manhours were saved; the newer aircraft still requires about twenty-four people per aircraft.[15]

---

[15]The Air Force's managers in its Tactical Air Forces (TACAIR) claim the increase in maintenance manhours has peaked. Now, presumably fewer people will be needed to handle the knowledge burden of the F15. (Binkin 1986:51–56). In 1986, TACAIR planned to reduce the flight-line specialities to only two by making each repairer system-specific (knowledgeable about all the functions on that model) rather than function-specific (knowledgeable about that function on all aircraft) (Binkin 1986:10). The difficulty is that TACAIR does not say what manpower is invested in the worldwide system tracking each of the F-15's forty-five black boxes individually. Manhours and costs removed from the front line are undoubtedly regained in this monitoring system (Demchak 1983). As noted

Other inaccurate assumptions also have had dramatic consequences. Estimates of mean times to repair (MTTR) were used in
calculations of necessary maintenance manhours. Included in the
MTTR were estimates of diagnostic times. For the M60A1 tank,
diagnosis was 7 percent of the total maintenance time; in an
electro-hydraulic tank, failures present a multitude of sensory
clues to make diagnosis easier. For the M1, however, operational
tests indicated that diagnostic time was 20 percent of the total
(Binkin 1986:64). During initial production with Chrysler, a failed
M1 tank stymied diagnosis by the manufacturer's experts for
nearly a week. A failure in the laser range-finder on the M1 also
stumped a manufacturer's representative; using the manual's
alternate test procedures, the engineer took three days to locate
the source (Demchak 1983). Other evidence indicated that diagnostic time for the M1 was underestimated in the managers'
planning. Despite the increase in diagnostic requirements for the
M1 over the M60, diagnosis was curtailed in the basic maintainer
classes for the M1. Even though diagnosis was time-consuming,
the number of maintainers on the M1 was reduced as well.[16]

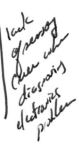

In addition to diagnostic time, the dramatic increase in the
number and size of manuals needed to test systems increased the
actual MTTR. The number of manuals, usually expressed in
pages, also dramatically increased, mushrooming to 27,000 pages
for the M1 and to three times that for the M60 (Bragg and
Demchak 1982:1, Nielsen 1983:10, Binkin 1986:44). The troubleshooting manuals for the M1 turret covered 3,000 pages and
included numerous references to other manuals (Demchak 1983).[17]

---

elsewhere, the maintenance problem does not go away when box is moved to the
depot (Spinney 1985:32-36).

[16]See Bragg and Demchak 1982. The estimates of MTTR for another system, the
Multiple Launcher Rocket System (MLRS), were also significantly underestimated.
Operational tests in 1983 showed an MTTR for the launcher's loader to be two
times the design standards. For the launcher itself, waiting for the right part,
tools, or maintainer reduced the availability of the machine by 20 percent (Binkin
1986:48, 65). These figures reflect a peacetime situation in which, especially in
tests, the parts and the personnel were more likely to be present; this does not
augur well for maintenance in wartime.

[17]It is necessary to note that in the M1 manuals the M1 tasks were more
specifically detailed than they were in the equivalent manuals for the M60A3. By
my inspection, however, the magnitude of the difference in pages was not fully
accounted for by this expansion of detail.

In one set of three manuals covering equivalent repair functions, the M1 manuals addressed 19,000 parts and the M60 manuals addressed 13,000. Costs per page were nearly doubled, as were words per page (Nielsen 1983:39).

Showing a widespread reluctance to use either test equipment or the associated manuals, maintainers were faced with the even more time-consuming but possibly more accurate alternate test procedures. Soldiers avoided using the STE-M1 because of the time it took to assemble it and the manuals, to climb in and out of the tank to reattach cables, and finally to read the exhaustive step-by-step diagnostic procedures (Demchak 1983, GAO 1981a:52, Gividen 1981:8). One common method of avoiding having to use cumbersome equipment was to replace boxes serially, but this raised the mean repair time in the field. In addition, any good parts removed were sent back into the repair system as failed parts, producing unnecessary scarcities throughout the system.

Other calculations have also been optimistic. The industry standard reliability measure—mean time between failure (MTBF) —is critical to estimates of maintenance manhours. If this figure is inaccurate, the structure of maintenance around a weapon system is seriously imbalanced. Surveys of electronics systems suggest that the laboratory estimates of MTBF need to be reduced by factors of 2.5 to 5.0 to reflect experiences in the field accurately. If this reduction is not made before personnel and equipment are distributed, parts procured, and people programmed, then a system said to have a ten-hour MTBF might experience only two to four hours of operation before failing (Nauta 1983:3-3). Such unexpectedly short hours of operation are deadly on a battlefield and catastrophic for a constrained supply and maintenance system.

For a tank, MTBF is represented as mean *miles* between failures (MMBF), and for the M1 the estimated MMBF was also seriously inaccurate. The standard was 101 MMBF, but during operational tests the tank achieved only 75 MMBF. This figure was adjusted by managerial-level representatives to 98 under assumptions of future improvements. After two years in the field, however, the tank had achieved only 66.8 MMBF—32 percent less than the adjusted figure (LEA 1979).

The consequences of poor test equipment reliability, scarcity of

[114]

skills, and faulty assumptions behind parts stockages were high rates of good parts wrongly removed from tanks. In the Army, these good parts mistakenly removed were labeled "No Evidence of Failure" (NEOF) by the support unit discovering them. NEOF rates indicate the level of inaccuracy of diagnosis among the units turning in bad parts. In 1981–82 more than 50 percent of the failed components sent from DS units to a combat electronics GS-level maintenance unit were good (NEOF) components; for four months in 1983 the figure was over 90 percent (Demchak 1983).[18]

In 1980 the M1 displayed a 56 percent NEOF rate during tests conducted with extensive contractor assistance (LEA 1979). Similarly, in 1983, tests of the MLRS exhibited a NEOF rate of 54 percent, as opposed to the required 7 percent (Binkin 1986:64). The numbers suggest a consistent tendency: when people were puzzled by broken equipment, they simply pulled components and sent them into the supply system. As a consequence, units gained less diagnostic experience and became more dependent on the supply system. The result was a drift of repairs rearward as diagnostic time limits and maintenance by "remove and replace" pushed difficult components back up the support chain.

These high NEOF rates also induced large budgets for parts among using units and created unanticipated scarcities in the supply system.[19] Because the maintainer often did not know which part was the bad part, all the removed parts were put in the system for repair. As with all electronic parts, no visual signs

---

[18]The current numbers should be compared with equivalent numbers from 1969 for tank-automotive fuel-electric components. The false removals were 40 percent of the "bad" components turned into GS units. The reason given at the time was lack of effective test equipment and was used as a justification for the mass acquisition of ATE (Brachman 1968:III-2).

[19]When good parts are inefficiently cycling through the support system, the costs to the organization increase. Research in 1975 on aircraft parts indicated that doubling the time required to get a new part, *ceteris paribus*, increased the cumulative costs of all parts involved by 58 percent (Smith 1975:17). Research in 1981 indicated that for each additional day a part is in the repair pipeline, a 1.0 to 1.5 percent increase in the overall spares procurement requirement is needed, just to maintain the weapons at existing levels of readiness (Hanks 1981:3). A 1 percent increase over the 500,000 line items managed by AMC would constitute an enormous increase in costs (Keith 1982:64). Such an increase would prompt serious cutbacks on other procurement and tighter control of parts usage by the depot commands.

[115]

of malfunction were available, so intervening levels of repair would just pass the removed parts back until the level authorized to break them open for repair was reached. The consequences of good parts being removed from the inventory exacerbated a pattern that was more and more typical of new equipment: there were difficult shortage problems due to budgetary limitations, overly optimistic assessments of parts requirements, and low equipment reliability in the field. Furthermore, repair by replacement induced failures in electronic machines by increasing the possibility of dust getting into critical nodes, pins getting bent, and highly precise calibrations being disturbed.

The consequences of faulty assumptions at the managerial level were imbalances and bottlenecks.[20] If the tank stood a good chance of failing if it was used, then not moving the tank avoided costly frequent failures. As a result, tanks in 1982–83 routinely spent twenty to twenty-five days a month in the motor pool. The low mileage, however, counts against the tank both in the MMBF and in the experience of the maintainers. One tank battalion commander had not moved his tanks in six weeks, and his chief maintainer refused to let any of the junior maintainers touch the tanks (Demchak 1983).[21]

There were other examples. In one M1 tank battalion in late 1982, five tanks had been deadlined for 353 days and heavily cannibalized, prompting a parts order considered excessive: 370 parts for only five tanks. Since these unusable tanks were the battalion's "float" tanks, intended for combat as well as for maintenance reserves, the maintenance responses were risky as well as costly. A tank battalion maintenance officer noted that, during a field exercise involving one brigade, all the other brigades were raided for their functional equipment so that the equipment would make it to the exercise area and back. Other investigators have made similar observations. For example, one

---

[20]The managers' assumptions about possible accuracy—i.e., no more than 10 to 20 percent NEOF at the next support level, seem quite low when the Army's own well-established IHAWK missile system has experienced 40 percent (Demchak 1983).

[21]Furthermore, the chief and his senior repairers did not use the STE. Instead he was in the process of developing his own mental library of failure pattern by climbing into the tank with a multimeter (Demchak 1983).

tank battalion reported losing four to five tanks en route to exercises, and another routinely returned from five-day exercises with 45 to 50 percent of its tanks not-mission-capable (NMC) (Demchak 1983).

Induced by high rates of mistaken removals (NEOF), high failure rates, and high costs in parts, supply imbalances emerged. Due to optimistic managerial level expectations,[22] one critical bottleneck in M1 support was a shortage of the essential M1 test equipment, the STE-M1. The STE-M1 was assigned one per company[23] for a total of six per M1 battalion; costs of the test equipment precluded a spare for each company. The allocation was low. Seven STE-M1s and three DSETS (ATE for the DS level) were needed to support only thirty-one M1s during OTIII—that is, two-thirds more equipment than the M1 battalion was authorized in order to support two-thirds the number of tanks in the normal battalion. This planned scarcity encouraged internal scheduling and monitoring between companies and the battalion. Furthermore, there were initially no authorized spare M1 engines or engine modules. By contrast, a fleet of 150 M60A1 tanks usually was supported by thirty serviceable spare engines (Demchak 1983).

As a result of the managers' underestimating key standards, critical specialities were kept at very low levels of skill in the tactical forces closest to the tank; crucial advanced levels of these specialities were intentionally concentrated, or migrated, to the rear. For example, the entry-level 34G fire control systems repairer in the battalion had no supervisor to provide the needed OJT; for the individual to receive the required education, the division direct support battalion had to provide the advanced expertise, encouraging the division to neglect the repairer or, more likely, move him or her rearward to the direct support battalion at the division level. Through the multiplicity of decisions by managers, the battalion never acquired the ballistic computer repair skills the

[22]Or parsimoniousness.

[23]"Lacking a serious analysis of test workload, the Army authorized one test set per company supporting 17 tanks. The consensus is that at least two sets are required to avoid waiting times, even ignoring the current support problems [software, test points, high level of skill required] of the STE itself" (Nauta 1983:4-19).

managers intended when they assigned such a speciality in the first place.

The shortcomings of the initial predictions and assumptions of the Army managers, as well as the inherent unknowns of the new machine, resulted in imbalances and bottlenecks and forced local, informal structural changes as reactions. The knowledge burden of the machine, especially its rogue outcomes, strained the allocations of TMDE, maintainer skills, and parts. In the tactical organization, pressures mounted to expand along the three indicators of complexity, moving from increased numbers through differentiation to interdependence.

In such constrained circumstances, an unintentional cycle develops in which individual decision-makers iterate reactively through options, each time increasing the complexity of the organization. The iteration begins with the easiest choice: obtain more of the critical item and use redundancy to accommodate the knowledge burden. The costs of fully meeting the shortfalls in knowledge, however, are prohibitive and unauthorized, so this option is limited. Some increase in numbers may occur, but it is usually less than requested. Each nonredundant piece of test equipment now becomes commensurately more critical because there are fewer than desired. A small increase in numbers occurs.

Pressure then builds to assure the availability of this equipment by giving it importance—specializing, for example. Having testers of the test equipment is a natural response and constitutes a small increase in differentiation. Mobility requirements and budgets, however, also constrain this option and greater job diversity is not enough.

Finally, to use the limited supply of the equipment or testers more efficiently, there is a tendency to have units share equipment or manuals. Interdependence increases, and units become more dependent on one another for essentials. Still, knowledge shortfalls persist. Iteratively passed along from numbers to differentiation to interdependence, the need for missing knowledge

[118]

forces a greater dependence on an outside source of knowledge—
that is, contractors or the supply system. This option, however, is
also constrained by the enormous costs and potential for disruption. The cycle repeats as the individuals involved return to
previous options, looking for ways to alleviate the shortfalls.

This cycle iterates as the limit on each indicator is reached.
When more people or more test equipment is not possible, then
more kinds of people or equipment are considered, and so on.
Each iteration tends spontaneously to produce slightly greater
numbers, differentiation, or interdependence, whichever indicator has any slack at the moment. As people in the technical core
adapt, or, via feedback mechanisms, managers react, the organization becomes more complex in response to the unaccommodated knowledge burden of the new machines. After the introduction
of the M1, a cycle of responsive changes iterated along the three
key indicators to produce a more complex tactical organization.

Increases in Numbers

Reactive changes increasing the number of personnel were not
formally possible because of the ceiling Congress put on Army
personnel. Especially in Europe, Congress left too few slots for
support units to get their full cadre of individuals. The number of
people required in order to maintain the new equipment was also
inhibited by the intense turnover of personnel in the field, especially in Europe. As a consequence of both the ceiling and the
turnover, units visited were short of people and expressed frustra-
tion at being unable to accommodate the new equipment with the
kind of attention required.[24]

[24]See Nauta 1983:3-9. Another system, the Patriot, offers an additional example
of how difficult it is to identify increases in numbers. The managers decided to
move Patriot maintenance from a three-level structure with intermediate maintenance, to a two-level structure in which the first level was the organizational-level
maintainer pulling a bad box and the next level was the depot repairing the box.
This structure presumably moved the personnel from the intermediate level in the
technical core to the depot level staffed by civilians. On the other hand, the
diagnosis problem was so severe, and presumably so expensive, that it was
necessary to create a highly skilled special team. On balance, it is difficult to say
whether the numbers increased or not, but it is clear that an unexpectedly large
knowledge burden forced the creation of the special team.

Increases in the numbers of equipment, however, faced no ceiling and emerged as supply or maintenance individuals tried to keep on hand more of critical components than was formally allowed by their PLL or ASL. The new parts have a much larger cost, which means that a unit could afford to ignore some parts and some it cannot ignore. A good example was a small plastic light panel on a Blackhawk which costs $553. When this item was broken, the unit was instructed to throw it away. For combat units, however, with about $15,000 a quarter to use for maintenance, throwing away a $553 item was enormously difficult for them to countenance and therefore, the maintenance personnel tended to save and try to repair items identified as discards (Demchak 1983).

The managers' rationale for identifying such an item as a discard was that it was more expensive than $553 to attempt to repair that item. For the tactical unit in the technical core, however, it was more expensive to buy that item from the supply system than to use time and trained personnel in an attempt to repair it. Another unit constructed its own test equipment, adding not only to the numbers but also to the differentiation of its equipment (Demchak 1983).

Increased Differentiation

Pressure on one fixed indicator often resulted in movement along another. Unable to acquire more people or equipment, units tended to specialize those they had—unable to increase numbers, they increased differentiation. For example, one battalion commander designated one of his scarce maintainers as an unauthorized parts-supply clerk—an increase not in numbers, as desired, but in differentiation (Demchak 1983).

Due to their generally unauthorized status, increases in differentiation were usually found within the unit. Desperate to keep control of critical maintenance supplies, some units informally differentiated internally, in contradiction to their authorized structure of organization.[25] They designated some individuals to track

[25]This structure is found in a unit's table of organization and equipment (TOE).

supply requests or maintenance problems full-time. These informal specializations included an unauthorized battalion maintenance officer instituted in the battalion having trouble with maintenance and several unauthorized supply clerks. A GS maintenance unit had unauthorized maintenance facilities specifically for dealing with the M1. A tank battalion that had just received the M1 planned to keep its quality-control section despite its upcoming elimination in the new unit structure (Demchak 1983).

Increases in the differentiation of parts and other support equipment were cumulatively profound, and the effects were multiple and wide-ranging. For example, parts have to be recognizable to be used. In the field, recognizing the parts was in itself a difficult task. A screening system was often established in order to get the correct parts. One unit had an unauthorized double-check system using a second clerk.[26] Issued a new maintenance supply computer, many units were supposed to maintain the older manual system as a backup in case of computer failure in wartime. Unfortunately, the new volume and diversity of parts made the older manual system too time-consuming to maintain, so a wartime backup system was often neglected, making the unit severely dependent on the one maintenance computer they were authorized.[27]

Such adaptations did not tend to disappear after the short-term imbalance was relieved. As the shortage in one component of one new piece of equipment eased, shortages in necessary components in other new pieces of equipment would emerge, and the reasons for this structural differentiation continued to be present. These adaptations also gave the commanders or staff officers an increased sense of control of the uncertainties of the supply and maintenance in their unit. By increasing the internal differentiation of the organization, however, these adaptations heightened

[26]In addition, in the early 1980s, private personal computers were emerging everywhere in the field. Unit members characterized their machines and the people dedicated to their use as simply helping the unit keep up with the new equipment.

[27]This observation is by no means an argument for a return to the manual system. It is cumbersome and inadequate for the volume of parts required, but it is an argument for redundancy in the computerized system, a redundancy that is both expensive and, if resilience is desired, unavoidable.

the potential significance of the loss of one of the differentiated components.

Increases in Interdependence

The most widespread of the responsive changes, interdependence, is the one indicator most likely to be affected by a large knowledge burden. In response to shortages of knowledge caused by shortcomings in ATE, training, and the managers' structural assumptions, the technical core units adapted spontaneously. While individual unit commanders preferred to be independent of other units for survival in wartime, they usually exhausted the options of increasing numbers and internal differentiation before they met the need for missing knowledge fully. As a result, unit leaders then gravitated toward the final and most expansive option, that of increasing their internal or external interdependence. They increased their dependence on such other sources of knowledge as test equipment or contractors, established screening activities, concentrated resources, and finally, avoided maintenance by transferring requirements to the supply system where possible.

Both a differentiation and an interdependence issue, automatic test equipment (ATE) works to increase the former via specialization and the latter via dependency. Since the M1 could not be diagnosed without the simplified test equipment (STE-M1), the unit with that test equipment was the only unit able to diagnose the weapon system.[28] Expensive equipment like the STE was

---

[28]The injunction that the M1 was not to be a complex tank was specifically stated by Congress and reiterated by the Army to the contractor (Demchak 1987: Appendix A). By incorporating modular components ("black boxes") into the tank, the development of the M1 was not intended to include development of test equipment outside of that which was built-in. However, the prototypes in 1976 revealed a level of internal complexity that the Army did not anticipate, and external test equipment, the STE-M1, was hurriedly ordered (Bragg and Demchak 1982). From then on, major weapon systems have been assumed to need external test equipment as well, and an STE-X family of test equipment has been under development. Because each item in the series is tailored to and can only be used on a particular weapon system, a subtle increase in differentiation as well as dependence occurs. With this proposed family of equipment, the awkward design shortcomings of the basic STE are perpetrated through its progeny, increasing the likelihood that future STEs will similarly fail to perform well.

[122]

allocated sparingly, inducing borrowing among units. The test equipment's relative scarcity, along with its criticality, induced changes in interdependence.

For example, a GS unit that was not authorized to have an STE-M1 and yet had to repair components of the M1 might have had to rely on another unit's expertise in using the STE and correctly identifying the source of the fault. Instead, however, the GS unit borrowed a tank from a combat unit to use as a giant piece of test equipment (called a hot mock-up). As loss of this tank would seriously damage maintenance operations, the unit was dependent on the combat unit that loaned the tank, but the borrowed tank could not provide the in-depth system the peculiar knowledge required to do the high level of repair this unit was intended to perform. As a result, the maintenance unit was gradually occupying its time by doing a lower level of maintenance that higher echelons presumed would be performed by the battalion lending the tank (Demchak 1983).

Differentiation induced increased interdependence, especially when the supply system was available as an alternative to maintenance. For example, one technical commodity command, the Army Armament and Material Readiness Command (DARCOM), allowed a DS unit to open certain black boxes and send the printed circuit cards back, while another command, Missile Command (MICOM), required the entire black box to be returned untouched to the depot for repair (Demchak 1983). Because some weapon systems have components from both commands, the DS unit needed to track the origin of a black box as well as its condition. The tendency to make unauthorized repairs on boxes ultimately resulted in another command, the Combat Electronics Command (CECOM) narrowly restricting permissible repairs and making the technical core more dependent on shipments from the United States. Another unit moved in the other direction, not opening black boxes even when authorized in a limited way; usually this neglect was rationalized by pointing to a lack of skill in the unit. The consequence of this differentiation was greater dependence on the supply system and on knowledge available in specialized units in the rear (Demchak 1983).

Contact teams constituted an interesting case of increasing

[123]

formal interdependence conflicting with informal arrangements. In the managers' anticipatory changes, contact teams were tailored to the requesting unit's needs, tying corps to particular divisions, and combat units directly to GS support units. The doctrinal structure reflects this distribution of skills by replacing expertise in combat and direct support units with the promise of expert contact teams. In practice, however, the multitude of types of equipment for which a DS maintenance battalion would be responsible forces internal specialization such that sending contact teams means sending the only specialist on that piece of equipment. Traditionally, the organizational maintenance level was responsible for bringing the equipment to the maintenance battalion, and that practice continued in the field despite the anticipatory changes in doctrine (Demchak 1983). As contact teams were infrequently sent out, the would-be requesting units turned to other sources of knowledge—contractors or representatives from the depots—and to the supply system.[29]

The increased use of contractors was an obvious increase in interdependence. Difficult maintenance was often directed to the most skilled personnel in the unit, or elsewhere outside the unit, so parts would not be wasted and criticism for nonworking machines could be deflected. Units consistently attempted to push maintenance onto the units behind them—presumably where more highly skilled personnel with better test equipment and larger parts supplies were located. As more troubleshooting tasks move to DS and from there to GS or depot, the company or battalion becomes less robust and more dependent on outside support for basic diagnosis (Demchak 1983).

Reliance on contractors distorted the managers' understanding of the true capabilities of the tactical units. For example, in the early 1980s, for major exercises like the annual NATO REFORGER, there were contact teams at each M1 battalion before and after the exercise. Furthermore, during the 1982 REFORGER exercise, there was a dedicated C5 transport flying from the contractor in Michi-

---

[29]The Army as an organization is more attentive to Defense Department restrictions on the use of contractors. The Navy, in contrast, keeps incomplete records and has difficulty identifying the extent of its dependence on contractors (Kaiser and Fabbro 1980).

gan to Frankfurt with M1 parts and engines. Hence, the battalions were not testing their capabilities in diagnosis and repair; rather, the knowledge was provided on-the-spot by contractors. According to interviewees, contractor representatives were doing maintenance in the field with their own specialized equipment during these major exercises in 1982. Army representatives later announced 90 percent availability for the M1 during the exercise (Demchak 1983).

Now officially permitted in the field during the first two years a new weapon system is deployed, Army contractors inadvertently encourage fragility in the system for wartime. With this policy, the nominal operational testing intended to take place before the equipment is accepted is in effect done after the equipment is in production. As the contractors discover problems, the improvement packages will be constrained by funding, leaving the maintainers in the field to accommodate knowledge left out due to expense. Furthermore, when the contractors leave in two years, if they do, the soldier-ATE combination will lack the knowledge available only through the contractors' skills. The relatively easier access to parts from the manufacturer and the relatively more skilled diagnosticians will also disappear. This loss will be expensive and frustrating in peacetime and deadly in wartime.

Scarcity of knowledge also induced other adaptations that have increased interdependence—particularly concentration of scarce resources, often to the rear. This rearrangement was most likely to occur within units. For example, direct support units that repair communications equipment have always had one person who was responsible for FM equipment and one person who did AM equipment. Resources formerly distributed across units, however, have been concentrated and located further back in the division area for monitoring or protection.

In one division, all the radio repairers were drawn out of the brigades and brought to the maintenance battalion in the division rear areas. One division commander withdrew all the spare tanks, meant to be maintenance floats in the divisions' reserves, back to the division headquarters in order to control the uncertainty in using the M1. The division commander had only five tanks to act as buffers for his division of 500 tanks; the original plan was for each battalion to have 5 tanks, for float for a total of about 45 tanks per division (Demchak 1983).[30] The result was increased fragility, because the only source of additional tanks in the division would be those five tanks distantly located in the division rear area.

Even as costly and difficult maintenance was avoided by levying the requirements on the supply system, the common practice of avoiding maintenance during exercises was also used by units to reduce the knowledge burden. In most units, all the possible and affordable maintenance—that is, replacement of boxes—was done before an exercise, and virtually no repair was done during the exercise in the field. After the exercise, it took weeks for the unit to recover from the deferred maintenance and that recovery depended on the budget and the supply system (Demchak 1983). The result was that, under most circumstances in peacetime, the technical core rarely actually tested its growing interdependence under approximate combat conditions.

Many of these increases in interdependence and in differentiation resulted from efforts to avoid paying the high costs of new parts. In particular, the establishment of parts-screening cells increased interdependence. If a front-line unit operated as structured in the formal documents, its use of parts was excessively expensive and its machines were often down. These elevated rates overloaded the supply system, producing machines waiting for parts and equally elevated outlays for parts and maintenance hours. The organizations supporting the front-line units then acted to control these rates. Intervening headquarters between the battalions and the depot routinely used their authority to cancel orders that headquarters judged to be inappropriate or exceeding funding levels.

---

[30]In 1968, when the 40 percent faulty removal rate was experienced on a much less complex tank, 2 tank engines were required in the pipeline for each ten tanks in operation, and 0.66 to 1.0 tank in the pipeline for each 2 operational tanks (Brachman 1968:III-2). A quick calculation based on a current division with six tank battalions with 54 tanks each suggests a need for about 115 to 174 spare tanks and 70 tank engines for each operational armor division. The figure is exorbitant but, given the faulty removal rates and overall reliability, perhaps not unreasonable in order to control the uncertainty.

Front-line unit maintenance and supply personnel characterized the cancellation of requests as unpredictable and described
many small measures they took to make the process less mysteri-
ous. Ordering too many parts was a frequent response to slow
deliveries and cancellations. One division-level maintenance unit
installed a separate supply-screening cell in an attempt to catch
any good parts before they were shipped farther back to the
manufacturer or depot. Before any repair could begin, all components returned for repair had to be put in the queue of the
screening cell to be tested (Demchak 1983). As shown in Figure 9,
a new and more time-consuming support structure was created
and constituted an increase in interdependence.

As a result of these monitoring, scheduling, restricting, and
concentrating mechanisms, scarcities in available money and stocks
occurred along with increases in the time required for completed
unit-to-unit transactions. For example, in 1982 it took two days for
a unit to get new components needed for the M60A3 in Fort
Hood. For the same division's M1s, whose parts could be obtained
only from the manufacturer or the depot, the turnaround time
ranged from two days to more than sixty days (Demchak 1983).

Interviews in summer 1983 indicated even longer turnaround
times for M1 parts and the key components of other weapon
systems. The common denominator among parts that were difficult to get was the expense of the part. This pattern was consistent with managerial-level decisions to stock fewer of the more
expensive parts.

Reliance on the supply system grew. M1 units developed close
relationships with technical advisers from the contracting office in
the depot or the support command. To avoid errors in parts
exchanges, M1 units in Europe in the early 1980s, for example,
were not permitted to exchange parts among themselves, a process called "cross-leveling." Instead, they ordered directly from the
depots. Consequently, parts were slow in arriving, disappeared if
they arrived at the wrong unit, and were regarded as generally
scarce. One front-line unit relied heavily on contractor representatives to make personal telephone calls to the depot in order to get
parts (Demchak 1983). The Army's managers also initiated a
spare-parts tracking system dedicated to the M1. The system was

[127]

Figure 9. Increases in structural interdependence

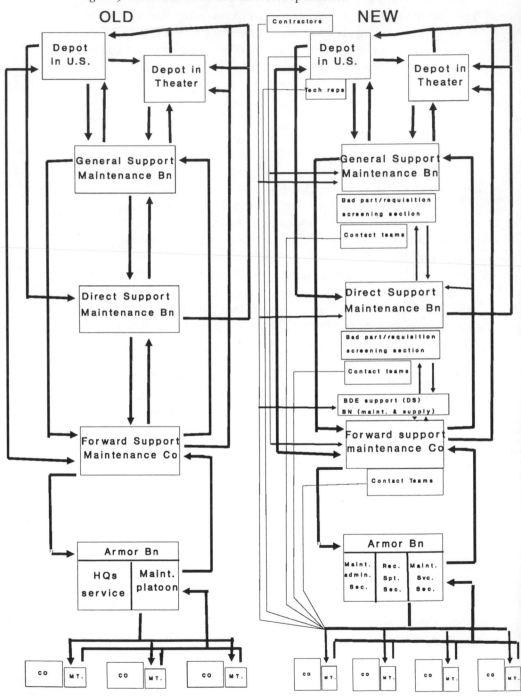

similar to though not as extensive as the system the Air Force used to track each F15 part (ALOG 1985a:40). The result was an M1 repair system that was heavily interdependent with the depots in the continental United States (Bragg and Demchak 1982).

In addition to the example of changes in tank units already given, there were other units with similar problems and special arrangements. An aviation unit borrowed parts from a unit not in their support chain in order to elevate the lender unit's stock requirements and increase the borrowing unit's chances of getting specialized parts it needed but was not allowed to order. The aviation repair unit had made informal arrangements with a nondivisional aviation repair unit to borrow parts on a regular basis. Another sign of this interdependence in an air defense artillery (ADA) battalion was a sergeant and a corporal who were assigned full-time to scrounging parts and were usually found in a truck running all over Europe looking for them. An aviation battalion deals directly with the program manager at the national inventory control point in order to have enough equipment to fly (Demchak 1983).

Scarcities of knowledge led to heavy reliance on the robustness of an intricate supply system for survival in wartime. Technical support ties between front-line units and depots located in the continental United States (CONUS) tightened. Air transportation and a presumed greater accuracy in parts and skill programming were substituted for inventory. In order to save the inventory costs of lengthy travel in a parts pipeline, Army managers began a program of direct deliveries of parts to units via an intricate supply system. In principle, the process operates much like an airmail letter: the requesting unit gets the part from the depot in the United States rather than from a depot in the unit's overseas theater or through the general support unit in the rear. Parts sustain maintenance in the technical core, and given the differentiation among the complex equipment, there are few alternatives to repairing a tank in wartime with other than the exact part. Hence, the tactical technical core was tied more tightly to the depots at home by these air deliveries; without a closer depot, their survival and accuracy becomes critical to forward maintenance.

Heavy reliance on the supply system raised interdependence to

[129]

a level that the institutional and managerial levels did not intend. It also produced subtle changes in priorities in the organization. In the traditional ordering of priorities, time matters more than money at the front lines, and money matters more than time at the depots. Somewhere between the two extremes there was a crossover point where both time and money were of equal importance. In previous major wars this crossover point was somewhere over the Atlantic. In the premodernization peacetime, the crossover point was located in the theater headquarters.

Given the high costs of parts and the burden of high NEOF (no evidence of failure) rates, this crossover point has slowly been moving closer to the front lines. It was definitely present in the direct-support level, where discussions about maintenance almost immediately turned to discussions about the cost of parts and about internal efforts to control their use. One executive officer for a tank battalion proudly pointed out that his commander had at least one unannounced movement of tanks a month. The unit was mobilized, went ten miles down the road, set up, and the next day tore everything down for the return trip. The reason for his pride was that, unlike other battalion commanders, this one was willing to exceed his budget to do this expensive training on a regular basis.[31]

MACHINE COMPLEXITY BEGETS ORGANIZATIONAL COMPLEXITY

The Army's managers set into motion forcing functions that produced a more complex, and more fragile, organization.[32] As the Defense Science Board stated in 1981, "Our high performance

[31]Although the M1 has a fuel consumption significantly larger than that of the M60A3, fuel was not perceived as the major problem in the tactical units. The cost of parts was consistently cited as the greatest problem, because of constant difficulty in diagnosing malfunctions. In part as a result of efforts to avoid using the tanks and consequently needing more parts, the tanks generally did not move enough to reach the limits of fuel budgets (Demchak 1983). See Tice 1988 for a discussion of this continuing problem.

[32]The Army's managers' "preoccupation with achieving maximum performance systems and fielding them as quickly as possible while attempting to contain costs has left little leeway to consider the implications for reliability and maintainability" (Binkin 1986:131).

systems have been force fit into maintenance and repair structures which preceded the advent of such systems, and . . . these structures were not well-matched to today's needs" (Binkin 1986:61).

As the last two chapters suggest, a more complex organization emerged in the tactical forces as a response to increased complexity in the M1 tank. With all the anticipatory changes and spontaneous adaptations, increases occurred in nearly all of the N, D, and I indicators to the extent permitted by the constraints. Pressures to increase in numbers enhanced pressures to increase differentiation. As differentiation was usually constrained by mobility requirements, these pressures then forced more interdependence to occur. Constrained by dispersion, mobility, and funds, increasingly interdependent tactical units came to rely heavily on the expertise and stability of the supply system. As the supply system proved incapable of meeting the costly needs of the maintenance units, pressures built again for increased differentiation and numbers. The cycle continued until, in principle, the knowledge burden became as accommodated as possible in the organization. The result was a more complex organization, the consequences of which are explored in the remainder of this book.

[7]

Surprise, Complexity, and Battle

Surprise is the special bane and boon of warriors. Historical experience has shown that, whatever the plan, operational realities will be disrupting and filled with surprises. In addition, any task that is tough in peacetime is even more difficult in wartime, because an enemy will deliberately try to disrupt the other side's operations. Bad weather; great fatigue among soldiers; long, slow, and vulnerable supply routes; a lack of information; and local shortages are all normal for an army, even in peacetime exercises. Wartime produces refugees, clogged routes, casualties, and fear.

All these events are foreseeable, but their frequency and magnitude are unpredictable; the universe of unknowns is exceptionally large. The enemy seems to appear suddenly or to unexpectedly win losing battles. During the battle for Antietam in September 1862, four divisions became logjammed around the crossing of a fordable river. As a result, Union general George McClellan decided against pressing the attack against Lee's forces. Unbeknown to McClellan, however, the Confederates were on the verge of collapse. Lack of information about either the river or the enemy caused a significantly inaccurate response. The American failure to foresee the German Ardennes offensive in 1944 is another example.

The large rogue set of war encourages confusion and disjointed responses to events, and the way this deadly unpredictability itself is seen has effects too. Infantry squads may avoid firing their weapons. Well-armed units in winnable battles collapse unexpectedly and catastrophically. In the fall of 1944, the 16,000 soldiers of the green 106th Division panicked and clogged the roads in flight; a massive traffic jam prevented reinforcements from arriving in time. In one case, only an officer with the rank of brigadier general was able to keep anything moving forward—he directed traffic at a key intersection.[1]

Surprise can also bring about victory for the side that is best able to use it on the enemy. Guerrillas consistently use surprise to disrupt government forces. The ability of the light, quick, and sturdy guerrilla to materialize seemingly out of nowhere is a consistent theme in the literature on insurgency. Losing governments are often slow and inaccurate in their response.

Many generations of military scholars have noticed the battlefield costs and advantages of surprise and the need for speed and accuracy in response. Sun Tsu, the eleventh-century Chinese military scholar allegedly responsible for many of the concepts behind Mao's military genius, urged military leaders to promote surprise, and hence disorder, in enemy forces. For Sun Tsu, 90 percent of war is won or lost before the actual battles began. The successful leader prepares extensively in advance in order to accurately gauge and quickly exploit the vulnerabilities of the opponent. Such planning not only increases the accuracy of the army's actions but also minimizes the length of the war, thereby also minimizing the opportunities for surprise in combat (Wing 1988). The 1991 Gulf War offers a successful example of avoiding surprise through extended, dedicated, and well-resourced planning.

Carl von Clausewitz also emphasized the unpredictability of war. Any plan would be overturned by wartime friction. "In war . . . the simplest thing is difficult. . . . Countless minor incidents . . . combine to lower the general level of performance (Clausewitz 1976:119). To compensate for the effects of friction, decisions should be made quickly and in the absence of clear information. The accuracy of the decisions would hinge on the commander's insight, based on his experience and practice within the army. For Clausewitz, peacetime experience helps the commander decide what is and is not possible, but this preparation has limits—there will always be friction.

---

[1]See, respectively, McPherson 1988, Marshall 1978:56, and Whiting 1981:54.

Many more recent scholars have reiterated the need for speed and accuracy in wartime, usually calling it flexibility or innovation: "Victory goes to the most flexible command structure" (Posen 1984:48). These advantages can come from faster, more accurate firearms or a more flexible technological system, but only when they enable a military force to act more quickly and accurately than its opponent under crucial conditions, including those of surprise.[2] The central concern of this chapter is the consequence for the newly complex Army organization of less accuracy and speed in wartime. How important are those abilities in the historical experiences of war, as well as in the way the Army currently plans to conduct a war? And how well do peacetime experiences provide useful information about possible responses to surprise in wartime?

ACCURATE AND SPEEDY RESPONSES

In the abstract, there are at least two basic requirements for an effective military operation: accuracy in problem and response identification, and speed in appropriate response. In wartime, survival begins with knowing the right response to unforeseen demands and being able to implement that response successfully and quickly.

⎨ *The More Complex, the Less Accurate*

Hypothesis 3: The more the complex a constrained organization and the more uncertain its environment, the less likely the organization will be able to respond rapidly and accurately to unpredictable environmental crises.

It is difficult for the complex tactical force to know with reasonable surety what happened and what needs to happen because there are so many nodes and sources of distorted information in its communications and command loops. The transmission of

---

[2]See, respectively, O'Connell 1989:190 and Van Creveld 1989:195.

information among a multitude of individual nodes, each of which must decide to make some adjustment to its position, invites misunderstandings and inappropriate adjustments. A nonmilitary example that is becoming classic in its retelling is that of Three Mile Island. Misunderstanding the information received, operators inadvertently drained water off the rapidly heating core when they should have flooded the container (Perrow 1984). And this accuracy problem occurred in a fixed facility without purposive enemy.

The history of war is replete with examples of accuracy errors. The attack on the *Liberty* and the loss of the *Pueblo* provide recent examples of this distortion. In the waning days of the 1967 Arab-Israeli War (the Six-Day War), a U.S. signals intelligence ship was diverted from its strategic listening mission for the National Security Agency (NSA) to a tactical mission by Navy officials. When NSA officials realized how close the ship was to shore, they prevailed on a reluctant Navy to call the ship away. Eleven hours later, the ship was devastated by an Israeli attack deliberately intended to stop the ship from listening to Israeli tactical communications. The message to move away from the shore had never arrived.

In the *Pueblo*'s case, a signals intelligence ship was sent to international waters off the northern coast of North Korea. Concerned about the tendency of North Korean fighters to ignore internationally recognized boundaries, a junior official at the NSA persuaded a senior official to send a warning to the Joint Chiefs of Staff (JCS) marked for "action." By the time the message traveled from the NSA through the JCS to the National Command Center to reach CINCPAC in Hawaii, its designation had changed to "for information." Some three weeks before the *Pueblo* was boarded by North Koreans, the message was read by some junior officers who filed and forgot it. The *Pueblo* never received the warning (Bamford 1982:283–287, 292–300).

The organization may not be able to acquire the correct information because of its own complexity. For the first years of the Iran-Iraq war of the 1980s, the Iranian military was unable to locate the multiple stockpiles of parts left from the spending binges of the deposed shah. The combination of widely dispersed

[135]

storage sites, independently operating military services, and a massive computer inventory system impenetrable to the uninitiated had hidden more than 30 million parts from the central military headquarters for over two and a half years of brutal warfare. During this period, the scarcity of parts made both tanks and aircraft unreliable weapons and contributed to a distinct passivity in Iranian armor and air units. Unable to use the modern equipment, Iranian military leaders turned to "human wave" tactics involving mass casualties.[3]

The lack of accuracy has its roots in decisions made before the war as well. By producing more complex tactical forces in peacetime, NATO would be more likely to face in wartime widespread requirements for improvisation, a wide variety of equipment, long lead times, and extensive parts shortages. The situation would develop like the German experience in World War II. Prewar German industrial production could not have supported the simultaneous production of both tanks and parts and so it emphasized tanks; as a result, a shortage of parts continued throughout the war. For the fighting units, extensive improvisations in repairs were necessary, especially during sustained combat; field repair shops attempted to repair all types of failures or battle damage. Although the official maintenance policy of the German Army said work requiring more than fourteen days would be evacuated to the depot, tanks were rarely shipped back for major repairs. Field shops kept tanks requiring extensive work and cannibalized them for parts. The parts shortage, the great distances between depot and front lines, and the constant transfers prohibitively lengthened the turnaround time if a tank was sent to the rear as planned before the war (Mueller-Hillebrand 1954:5, 18–26, 41).

The More Complex, the Less Speedy

The more complex organization must devote more time to coordination of actions. More nodes must be apprised of changes

[3]See O'Ballance 1988:103. The civil unrest in Iran and the distrust of the regular military by the revolutionaries were critical background factors to these tactical

that might affect the organization's internal relations. Over time, standard routing of messages evolves to give key nodes a chance to respond and protect their inputs or position if they deem fit. The result is that messages travel slowly and all too often are distorted en route. The *Pueblo* and the *Liberty* examples illustrate the problem of too little speed as well as accuracy. When the correct information was acquired, the system was too slow in transmitting it to warn the affected forces.

The classic example, of course, is the message intended to warn the American fleet in Pearl Harbor of the Japanese ultimatum. The message lay in a Western Union office in Hawaii—too late to be of use because of complex security requirements. If these rules had been ignored, the information still would have been directed to the Philippines and Panama first and then much later to Hawaii (Layton 1985:305–6).

An inability to coordinate responses rapidly, no matter how accurate the intended changes, has severely damaged numerous potentially successful operations. In the 1973 Yom Kippur War, the Egyptians failed to exploit the advantage of their surprise and push inland largely because of the time-consuming coordination required to get the necessary air support (Dupuy 1978, Middleton 1982). During the Falklands War of 1982, the Argentine military services could not coordinate their attacks or responses to attacks as needed in order to use their modern weaponry effectively. The services had never practiced with their advanced systems and, apart from the animosity among senior leaders, were unprepared to move quickly and responsively to crises (Hastings and Jenkins 1983:323–25). This situation emerged in World War II, especially among British ground forces, as well as in the Iranian forces in the 1980s Iran-Iraq war (Millett and Murray 1989, O'Ballance 1988).

In any war, then, a complex organization will find it difficult and expensive to respond to the surprises it faces. Across theaters, time, and technologies, complexity imposes surprise and confounds initiative. "Doctrine," however, is the military organization's antidote to surprise—if it is implemented and if it works.

---

conditions, but revolutionaries would have used the equipment if it had been available.

AirLand Battle Doctrine and Surprise

"Doctrine" is the military service's formal expression of its best guesses about what actions will prevail in war.[4] It is the set of "principles, policies and concepts [that govern] all components of a military force in combat and [it assures] consistent, coordinated employment of these components" (Dupuy 1980:9). Doctrine sets out a procedure that guides the rest of the organization in its tactics, other operations, and subgoals. The choices in doctrine are greatly influenced by the culture and common presumptions of the military service.

If doctrine is to guide the organization to success, it must blend well with the forces using the new complex weapons and with the combat axioms of history. If the doctrine cannot be executed by the organization, the probability of nasty surprises increases greatly. In a major theater of war, more complex tactical forces conflict with critical assumptions behind the U.S. Army's battle doctrine: the AirLand Battle Doctrine.

Relying on Speed and Accuracy

The U.S. Army presents its best guess about war-fighting with the overall doctrine published in FM 100-5.[5] In the three versions

[4]There is a continuing debate about the sources, changes, and efforts of doctrine (see Posen 1984 and Rosen 1988 for recent contributions). I am using a highly functionalized notion of doctrine as an organizational communication device that transmits the organization's overall current best guess about the best way to fight a war. This usage is tangential to all those debates.

[5]Williamson Murray argues that doctrine is largely ignored by military officers both in war and in selecting weapons. While I agree that the specifics of a doctrine are not closely followed, the doctrines' underlying themes do serve to coordinate roughly the loosely linked decisions of a large organization. They anchor the expectations of the senior leaders, aggregate roughly the weapons decisions, and provide a common "sheet of music" for the various functions in the organization. As others have noted, doctrine is valuable for its statements of what the organization *wants* to do, as opposed to what it *can* do. In this sense, it is justifiable to hold the organization accountable for discrepancies between its weapons or organizational decisions and its doctrinal intentions (Van Creveld 1982:178, Posen 1984:14). As Van Creveld (1989:178) said, "[written] records may not give a true picture of what an army was, but they do at any rate demonstrate what it wanted to be. Probably few soldiers read, much less digested, the bellicose reflections on the nature of war contained in the *Truppen Fuhrung* of 1936...yet the view put forward constituted an ideal to be striven for which, directly and indirectly, affected practice in a thousand ways."

of this document encompassing the Army's modernization efforts (1976–86), the Army has declared that speed and accuracy are to be used as substitutes for initial scarcity in materiel, soldiers, and advance warning in an overseas theater—and thereby to win even though outnumbered.[6] Called "AirLand Battle," this doctrine states that the battles of the 1980s and 1990s will be battles in which "offensives are *rapid*, violent operations that . . . *exploit success promptly*" (FM 100-5 1982:7-2, emphasis added).

Success in AirLand Battle is directly tied to the achievement of speed and accuracy. "Combat power is useless unless it can be brought to bear quickly—at the right place and at the right time." Accuracy in responses was equally critical. As the 1976 FM 100-5 stated, "Outnumbered forces cannot afford mistakes." Other publications elaborating on this doctrine are peppered with phrases like "see deep and begin early," "move fast," and "strike quickly." This emphasis permeates the document.[7]

This basic reliance on speed and accuracy is an understandable response to constrained conditions. The doctrine was created when the most likely place for a major conventional war involving the U.S. Army fully was the European theater.[8] In this scenario the United States and its allies would have begun the war decidedly outnumbered: 84 NATO divisions to 173 Warsaw Pact divisions.[9] Furthermore, these forces did not have much depth. In Europe, allied forces were about one corps deep all along the front lines.

[6]If the needed material is not anywhere in the organization's stocks and is not obtainable and there is no substitute, such speed and accuracy will not compensate for redundancy. For example, all the will and reorganization in the world will not provide an extra piece of test equipment if only three pieces are available and four are needed.

[7]See, respectively, FM 100-5 (1976:2-26, 3-6), TRADOC PAM 525-25 (1981:8), and FM 100-5 (1982:8-7).

[8]Many experts do not think such a scenario is likely, especially with Gorbachev's efforts to reduce military tensions. For a widely diverse European discussion of this viewpoint, see the alternative defense literature and works by such authors as Horst Ahfeldt and Lutz Unterseher. A good source in English for this literature is the journal *International Security*. It is useful, however, to remember the uniqueness of the last forty years of bi-polarity. As the international system becomes increasingly multipolar, the next forty years might not be as stable (Waltz 1979).

[9]The estimates of Warsaw Pact numerical superiority in tanks ranged up to five Soviet tanks to each NATO tank, depending on how Soviet reserves and Soviet allies were counted. In any case, there was little depth to the NATO forces; there were too few NATO combat units to keep in the rear for an operational reserve (Gessert 1983:25).

Because a corps' fighting capability is usually only one division deep, the available defense at the outset of the war would have been only one division deep. Such a major land war is the most stringent and demanding test of doctrine. In World War II, twenty-two U.S. divisions were involved in the Normandy assault itself. In the area defended in 1989 by six divisions in two corps, fifty-seven U.S. divisions fought in 1945 (Coakley and Leighton 1968:371). Depth is in essence redundancy,[10] and if redundancy is not available, forces and material have to be targeted more accurately and quickly in order to compensate.[11]

In the AirLand Battle doctrine, the rapid and accurate use of limited resources in space and time is to compensate for the lack in redundancy of materiel. The U.S. forces must survive from existing stocks while waiting for the home industrial base to provide replacement people, machines, and units. Military managers need to know where the limited material would do the most good and move it there in time. (FM 100-5, 1986:16). In setting these goals, the AirLand battle doctrine authors acknowledged the unknowns of war, but they intended to meet them by more precisely combining organizations and equipment meant to operate quickly and accurately in most circumstances.

With a focus on accuracy and speed, AirLand Battle as a doctrine is highly knowledge-intensive, emphasizing the preemptive "deep attack." It is intended to "extend the battle" forward in space (into the enemy's rear areas), deep in time (using planning

[10]Landau (1969) suggests organizational redundancy as an antidote to complexity. Logic and the U.S. experience of World War II suggest that uncertainty may be ameliorated by having more of everything. However, redundancy for a military is generally limited by the pressing need to be mobile as well as by the enormous costs of such an inventory. Tactical forces cannot easily achieve rapid and accurate responses through sheer massive duplication of units, and there is only so much they can dismantle, carry, and/or reconstruct rapidly. During the 1991 Gulf War, 84 percent of U.S. forces were outside of the theater providing enormous redundancy in in-theater assests. Despite this wealth, an unhindered six months of trial-and-error learning were required to prepare the organization for successful operations. The problems of complexity are neither cheaply nor quickly solved. See DOD 1990:II-5 and "Forces in Place" 1991.

[11]Shortages of time and ammunition are often used to justify the U.S. Army's strong interest in accuracy in firing weapons. Hence, organizational efforts to improve targeting and mass attacks are really expressions of efforts to improve accuracy. See Fallows (1981) for a discussion of the M16 and accuracy.

as backward mapping to give much advanced notice), and higher in other military assets (using such national-level resources as satellites and the Air Force for advance warning). The doctrine emphasizes the use of surveillance systems, accurate delivery of firepower, and near real-time communications in order to be able to maneuver the combat units appropriately (TRADOC PAM 525-25 1981:3, 7).

It is also material-intensive—a "fluid, destructive" battle is "resource-hungry." The "center of gravity of one or both combatants will be found in their support structures." Broken equipment will be key assets because "good maintenance practices in all units, forward positioning of maintenance units, stocks of repair parts and replacement equipment, and well understood priorities for recovery and repair may spell the difference between tactical success or failure."[12]

Expressing the underlying requirements for speed and accuracy, FM 100-5 lists requirements and implementation goals intended to guide the organization to producing the necessary quick and correct responses. The 1986 version of FM 100-5 identifies five fundamental requirements for success: anticipation, integration, continuity, responsiveness, and improvisation.[13] These are the outputs of the organization which would produce success on the battlefield if correctly implemented, and they are essentially tests for the organization constructed by its planners.

If the organization is to succeed using this doctrine, it will have to meet these five requirements in wartime. FM 100-5 and related documents were clearly written by committee and contain a confusing hierarchy of goals, subgoals, and tenets. As a whole,

---

[12]See, respectively, FM 100-5 (1986:7, 59, 61) and TRADOC PAM 525-12 (1981:6). As Clausewitz (1976:131) stated, "nothing is more common than to find considerations of supply affecting the strategic lines of a campaign and a war."

[13]In a military setting, the term "requirements" differs significantly from the civilian understanding of the word. It means inputs for success, not inputs for the organization. As a result, in a military setting and to a military commander, "requirements" do not mean something a commander can demand from the organization; they mean something the organization will demand that he or she provide as an output so that the whole organization's output produces (becomes "input" for) success. Requirements applied to the organization's actions—as these are—are in reality tests for the organization itself. This subtle but highly consequential twist in use often makes reading military documents confusing for the uninitiated.

however, these various lists endorse the need for speed and accuracy in military reactions to operational surprises. For example, the first requirement, *anticipation*, is the ability to foresee future operations and demands, both friendly and enemy, as accurately as possible.[14] In essence, this requires accurate and timely information and the ability to move the appropriate forces to the right places quickly.

The second requirement, *integration*, states that operations are to be "supportable at every stage" of a plan's execution (FM 100-5 1986:62). Integration is the coordination of all combat arms as well as support arms; accuracy and speed in the flows of organizational information are therefore essential. The third requirement, *continuity*, is avoidance of interruptions in operations. The forces must "never become hostage to a single line or mode of transport" (FM 100-5 1986:63). Continuity of operations depends on accurate, quick responses that overcome disruptions. The fourth requirement, *responsiveness*, demands that support units be "trained to respond on short notice and to 'surge' their support for brief periods" (ibid.). As the battle alters, the tactical forces must be able to adapt quickly and accurately and to operate successfully.

Drawing in part on all of the previous four requirements the final requirement, *improvisation*, is possibly the most important response to the friction of war. As FM 100-5 (1986) explicitly states, "improvisation has long been one of the American soldier's greatest strengths and should be viewed as an advantage in meeting emergencies," and as Janowitz says, "The increased mechanization of warfare often calls for more, rather than less, on-the-spot improvisation."[15] To survive, tactical forces will de-

---

[14]See FM 100-5 1986:62–63. Relying heavily on accurate information, the "Deep Attack" part of the battle, for example, occurs mostly by air when the enemy is twelve to seventy-two hours away from reaching the front of friendly troops. The tools of a deep attack are "interdiction—air, artillery, special operating forces, offensive electronic warfare, and deception." When the enemy reaches twelve hours away, the military focus switches from air to land battles (TRADOC PAM 525-25 1981:10).

[15]See, respectively, FM 100-5 1986:63 and Janowitz 1971:26. "After the Allies failed to break out of Normandy, infantry divisions were retrained in the narrow bridgehead literally hours before they were committed to battle, while ordnance teams *redesigned* new tanks to dig through the brush" (Janowitz 1971:26, emphasis added).

pend on their ability *to accurately develop expedient responses* in equipment or organization to the changes in tactical conditions.

The AirLand Battle doctrine, as stated in FM 100-5, therefore requires speed and accuracy in responses for its successful implementation. Adaptations that inhibit either or both of these characteristics fail to meet the requirements of the doctrine. To the extent that this doctrine could have prevailed in a future large-scale war, failing to meet its requirements means the organization would face such a war with a doctrine it could not use effectively.

Ironically, the doctrine's tenets encouraged the selection of equipment that induced greater complexity in the tactical forces and endangered the very speed and accuracy needed. In principle, the Army's doctrine is the guiding document for "fielding, improving and modifying the Army['s] . . . organizations, systems and equipment through the remainder of the 1980's and into the 1990's." It is "the keystone of force modernization" (TRADOC PAM 525-25 1981:24). Equipment purchased for the troops is chosen according to whether it was consistent with the doctrine and hence with the need for speed and accuracy. The Army chief of staff stated in a 1981 speech: "We need advantages in the equipment itself and in the means and methods of its employment. The gunners need a second or two advantage. . . . The battalion needs minutes. . . . The senior commanders need hours of advantage to . . . assure our strength at the decisive point."[16]

AirLand Battle equipment was to maximize these seconds, minutes, and hours while enabling the organization to anticipate, integrate, continue, respond, and improvise. Instead, a more complex organization emerged in response to the machines, hindering both accuracy and speed, especially in a critical arena—maintenance. Rippled through the rest of the tactical forces, the effects have endangered both implementation of the doctrine and the effectiveness of the forces in meeting surprises.

[16]See, respectively, TRADOC PAM 525-25 (1981:1) and Meyer 1984:198. Note that because FM 100-5 does not specifically use the label "AirLand Battle" the relationship between AirLand Battle and FM 100-5 is sometime confused. The clearest way to view the connection is to understand FM 100-5 as stating the fundamentals that AirLand Battle then arranges into more concrete forms. In principle, those fundamentals could be rearranged into another set of forms that could be called something else in the future.

Problems with Accuracy

Although one requirement addressed by the AirLand Battle doctrine is anticipation of future needs, the Army's managers have consistently had serious problems with accuracy, especially when predicting the actual performance of the machines for which the force's structure was optimized. Among other things, high rates of mistaken removals are evidence of these shortcomings. The less often anticipations are correct, the greater the need for elaboration and improvisation. By December 1984 at least twenty-nine hotlines—some on maintenance, some on specific weapon systems—had been established to provide guidance and knowledge when the anticipated training and aids were not sufficient (ALOG, 1984a:44–45). The peacetime need for nearly three dozen specialized sources of information suggests a telling dependency, but in wartime this type of support is unlikely. These hotlines are dependent on established peacetime telephone networks that are unlikely to survive in a major theater or to be present in less developed areas.

Accurately anticipating the wartime capabilities of ground forces in peacetime is particularly difficult because the exercises tend to lack realism. Integration, the second of the warfighting requirements, is virtually untested in realistic-size units with the new machines. Planners developed a new division structure, DIV 86, which was meant to optimize weapon system capabilities, but it was tested only with battalions at the National Training Center (NTC) in California. There are likely to be serious gaps in the wartime support capabilities behind these battalions. The support units would be closely linked to one another for parts and knowledge, and the mobile support of contact diagnostic or repair teams is unlikely to be available. In the violent, fast-paced operations of the AirLand Battle, "support units have thin-skinned vehicles, cumbersome machinery, limited weaponry and communications, and bulky stocks" (Reiss and Lee 1986:25).

The complexity of the new machines strains the peacetime structure often enough that there is a pronounced disinclination really to test the machines during exercises. The unit that does so will pay heavily after the exercise in maintenance hours and parts.

[144]

As a consequence, maintenance is not performed, armament not fired, resupply not disrupted, logistics communication nets not jammed, and critical sensitive surveillance or intelligence systems not used. It is also important to note that because the exercises do not actually destroy equipment, battle-damage repairs are not realistically tested. Planned expedient repair on battle-damaged vehicles may not be possible because newer machines continue to be precision-crafted and do not work well with quick fixes.

Because of the lack of valid exercise experience, reliable information about a lack of speed or accuracy may not even be recognized as valid in wartime. Subtle signs of impending collapse are more likely to be viewed as anomalies when peacetime exercises do not include certain difficulties. For example, the increased dependence or resupply means that one of the easiest ways to affect a combat brigade's operations will be to disrupt the links to its maintenance support or the maintenance units themselves, but this eventuality tends not to be realistically exercised.

Despite the Army managers' efforts to reduce maintenance in combat areas, maintenance will be critical in the next major war. During the conduct of the real battle, however, operational decisions will include hasty "relocation of support bases, major redirection of supply flows, reallocation of transportation means or short notice transfer of units from one part of the theater to another." If high rates of mistaken removals (NEOF), high parts costs, low inventories, limited technical knowledge and widespread sharing or scrounging occur in peacetime, wartime conditions cannot avoid making things worse for speed and accuracy in maintenance actions.[17]

An example of the intersection of exercise logic and higher headquarters inexperience might be the following hypothetical scenario. Greatly influenced by peacetime experience with flank positioning, a division commander significantly reduces the maintenance priority of a brigade on the flank of an attack. The

[17]See, respectively, Cavanaugh 1983:10 and FM 100-5 1986:63. Brewer argues that, in organizations, "consensus on matters of fact and on recommendations for action decreases with increasing size" (Brewer 1975:204). The situation worsens when obtaining facts regardless of disputes over their validity is as difficult as the verification itself. The inability of senior commanders to get information about activities at lower levels is a common phenomenon of wartime.

commander assumes that the level of resource consumption experienced in peacetime exercises will roughly allow the flanking unit's tanks to continue to operate. In an exercise, however, a flanking unit commander knows there is not really anybody out there and is not vigorous in patrolling or consuming supplies. Faced with the possibility of actual combat, however, the flank brigade commander aggressively maintains a high level of operation, and tanks fail more often, leaving the flank with fewer critical weapons. The division commander is then faced with a serious dilemma: pulling support from the main point of operations—risking success there—to shore up the flank, or permitting weakness at a traditionally dangerous location.

The British and American navies of pre–World War I displayed extraordinary abilities to pursue exercise logic to its extreme. In part because senior leaders developed their skills using flat, two-dimensional board games and surface ships in the late 1890s and early 1900s, both services thought in terms of the surface dimension of the seas. They largely refused to credit the subsurface threat from torpedo-laden submarines and mines and the aerial threat from planes. When war came and the true subsurface threat became abundantly clear, the grand fleets of the world hesitated to give chase for fear the prey was leading the battleships over submerged submarines and mines. Consequently, there was a general timidity and unwillingness to give battle as planned before the war (O'Connell 1989:212–31). In the 1980s and 1990s such blind spots are called vector logic. Unable to continuously monitor and respond to a 360-degree perimeter, carrier battle groups often prepare to fight in certain attack vectors chosen according to assumptions about the enemy and the limitations of ship equipment. Enemies who do not kindly arrange themselves in the correct vector of approach will have a field day.

In efforts to anticipate wartime needs accurately, the U.S. Army has recognized the problem of "exercise logic" at the combat battalion level and attempted to build an environment of realistic systemic stress at the National Training Center (NTC) at Fort Irwin, California. By most accounts, NTC personnel successfully stress battalion task forces one at a time. A brigade headquarters participates in the exercise, but it controls only the forces of one combat battalion in a task force. The M1 tank performs well at the NTC, but the preparations for the exercise are intense and dedicated to passing the tests. Much maintenance is deferred at the battalion level until after the exercise. In wartime, many more units will be involved. The higher-level headquarters will have to cope with the complexity of supporting the M1 and other new machines, but these staffs tend not to be participants at the NTC.

Integration has also been poorly tested at the boundary between the division and corps level of the organization. Over the period 1980–87, the dependence on maintenance units deepened while the division structure was being lightened, mostly in its support assets, in order to conform to congressional interest in lean formations. In ordering the Army's managers to reduce the size of these divisions, the Army's senior leadership was responding to a peacetime imperative that contradicts the likely wartime requirements of the new machines. The new light divisions, in particularly, have difficulty operating for long without corps support. Division operations in the future are likely to be less successful without significant levels of support from the corps maintenance units.

The potential inaccuracies of exercise logic applies to more than the tank itself. The success of this doctrine is also heavily dependent on the operation of a wide array of today's advanced machines to provide accurate information on what, where, when, and how to strike, how hard, how frequently, and how successfully. Exceptionally difficult to test in peacetime, the "Deep Attack" itself depends on many complex systems required to support it successfully. Surprises here make it unlikely the deep attacks will provide the intelligence, and hence the delay, required for effective ground defense.

Prior wartime experiences can be as misleading as exercise results in the search for accuracy in operations, especially in having the right part at the right moment. The Army's most successful major war experience occurred with troops in the offense, not the defense. During World War II, in the U.S. Army's race across France, maintenance companies without parts cannibalized extensively, but they had the most luck cannibalizing vehicles broken down by "hard wear in pursuit," not by battle

[147]

damage (Mayo 1968:277). During the 1991 Gulf War, the lack of a serious opposition made expedient battle damage repair irrelevant. Of 1956 M1 tanks in theater, only eight were damaged in combat operations (UPI:1991). In the defense, however, with troops holding ground or giving way, battle-damaged equipment should outweigh worn equipment, leaving fewer opportunities for cannibalization of scarce working parts. Furthermore, in peacetime, actually destroying equipment in order to practice repairing it is extremely expensive, and practicing battle damage repair in peacetime can induce failures in the equipment. This practice has been prohibited in the peacetime Army for all but the most simple of expedients. Even if practiced in peacetime, repairs with string and soda cans will not help much when the electronics are damaged.

The doctrine in essence also holds the accuracy of ground operations hostage to the robustness of another complex organization. It relies heavily on the assets of the U.S. Air Force to survive. Air Force equipment and crews are to give Army forces advance intelligence and to attack enemy support structures, thereby disrupting enemy momentum. Air superiority, or at least parity, is critical to air operations gathering intelligence or interdicting supply lines. Superiority is highly unlikely in a European theater, and air parity may be tenuous until well into the war.[18] Furthermore, as noted throughout this book, Air Force equipment is as complex and as prone to surprise as other complex systems.[19] With ineffective deep attacks, ground commanders will not get the information or extra time needed; their units will face the full brunt of the attack on the enemy's schedule.[20] This reliance on a

---

[18]The Air Force may not necessarily be willing to participate in the Deep Attack. It is unlikely to be eager to send many of its aircraft into enemy-held areas where they would have no numerical advantage and a vulnerable flight home. The Air Force would face risking assets where surviving pilots are not recoverable; the attacks would not benefit the Air Force directly by targeting airfields to reduce the enemy aircraft assets. If the Deep Attack is not successful, the Air Force would lose critical assets in vain and would have faced the major air battles over Western Europe short of firepower.

[19]See my comments about the F15/16 in Chapters 4 and 5.

[20]At least one document acknowledges this possibility. "We would like the deep attack to destroy enemy forces before they enter the close-in battle, but in today's terms and in all probability, tomorrow's as well, expense and scarcity of assets will limit the practically achievable effects of [deep attack in] delay and disruption" (TRADOC PAM 525-25 1981:15, 19).

sister service itself struggling to survive is likely to be costly. /

The fifth warfighting requirement, improvisation, can be successful only if the improviser is able to come up with and implement a good substitute accurately. There are two levels to this issue: the complexity of the machine and the complexity of the organization. Complex machines permit little improvisation in maintenance. Too few and too specialized printed circuit boards, electronics that are too integrated, too little training, and too much dependence on other units limit the number of improvisations that succeed. It is more difficult to use a paperclip to fix the highly precise circuit boards. As you lose black boxes, you lose significant combat capability. During the early years of the 1980s Iran-Iraq War, Iran attempted to improvise the manufacture of high-precision spare parts. As a result, a large portion of Iranian aircraft losses were mechanical failure accidents rather than combat losses, and the Iranian Air Force's contribution to the war was negligible (O'Ballance 1988:123).

There are limits to creative answers to machine surprises. During World War II, German industry tried to develop better tanks, but the increasing variety of new tanks exacerbated the wartime shortages. In the autumn of 1942, the new development Tiger tank was introduced into the Russian front with one spare engine and one spare transmission for each ten tanks; shortly after, nearly all these tanks were deadlined for lack of parts. The equally new Panther tank had the same experience. Given the uniqueness and scarcity of its parts, improvisation was not possible, and the fleet of new tanks was deadlined.

Organizational improvisation is difficult to direct accurately. Common but disruptive wartime examples include informal pooling of parts, parts raids on other units, violation of combat cannibalization, hijacking of experts, and radio failures at opportune times when contact teams are forward. Such forms of improvisation are more likely to happen and more likely to be severely dysfunctional when parts and experts are scarce. Widespread improvisations across a theater can severely disrupt operations and the allocation of scarce resources. Again, an example from the German experience in World War II is useful. German maintenance companies began sending squads to the in-theater

[149]

depots basically in order to hijack whole shipments of parts off the arriving German supply trains, whether the supplies were intended for the hijacking unit or not. Company commanders used bribery to make personal purchases of parts from industrial firms. More recently, in Vietnam, armored vehicle spare parts were available only for 50 to 60 percent of the demands. The rest were obtained only by cannibalization and scrounging, despite the affluence of American industry and the small size of the conflict.[21]

One unit's innovation may disrupt another unit's plan of action. For example, to respond to the deepening mud in the fields of France in 1944, First Army ammunition units set up their ammunition points on the sides of roads. This adaptation "strung out the ASP's [ammunition supply points], increasing the total mileage and hampered operations in areas far forward when tactical units had to use the roads... it took an armored division two days to pass through" (Mayo 1968:280).

The sheer number of concerned activities in an organization also inhibit improvisations. For large organizations, the speed at which material and information flows is a function of the ability of the system to control or eliminate the unexpected. If requisitions that do not meet presentational or unit qualification requirements are rejected in peacetime, they will be rejected in wartime too. Controls initiated to control costs in peacetime will tend to continue to control flow in wartime. In Vietnam the number of multiplying headquarters is said to have produced severe distortions. It took seven months for the headquarters conducting the operation to plan and implement a rescue of the prisoners at Son Toy in 1970 (Van Creveld 1985:249). To the extent, then, that peacetime arrangements meant to control deviations in operations serve to make organizational interactions less flexible and more complex, they directly hamper accuracy and key requirements of the AirLand Battle doctrine: anticipation, improvisation, and responsiveness in wartime.

[21]See, respectively, Mueller-Hillebrand 1954:5, 18–26, 41 and Starry 1978:224. In World War II, late in 1944, whole German convoys would occasionally be hijacked by forward units. When arriving at the designated drop-off point, the local commander would not allow trucks to unload. Instead, the entire convoy was moved ahead for days as a sort of mobile dump. Such improvisation by local commanders could be fatal to the larger plan (Bykofsky and Larson 1957:329).

Problems with Rapid Response

Army preparations for war do not auger well for the Army's ability to respond rapidly to crises in a major theater of war as the doctrine requires. The third requirement, continuity, is in essence a demand for supply redundancy which conflicts with peacetime practices. Increased speed in support during peacetime is based upon aerial resupply, especially of repair parts, from depots in the United States.[22] This aerial system's vulnerability to enemy air forces is, of course, enormous. A proposed in-theater depot will have to provide key backup repair parts to meet fluctuations; whether this depot would be adequately stocked is as open to question as the other estimates of wartime needs. Its inventory too is based on truncated exercise experiences and the demand practices of peacetime.[23] For example, in September 1944, the First Army received only 10 percent of the spare parts and assemblies it needed. Adding to the general friction of war were "bad guesses in the United States as to battle losses in tanks and the number of trucks and all kinds of vehicular spare parts that would be needed" (Mayo 1968:296).

Intended for quick initial back-up, reserves of materiel already in the theater are constrained by costs in both peacetime and wartime. A corps' reserve of tanks and other major equipment held in the in-theater depot is limited to one day of battle losses (TRADOC PAM 525-12 1981:4). Because the tanks and reserves of lower-level units would be fully occupied in each respective unit's area and because the only feasible resupply would be slow sea transport, the one-day resupply rule is likely to be too slow for

[22]In a seminar given at the U.S. Military Academy in the spring of 1990, a senior Army officer reiterated the Army's intention to use air delivery during wartime by saying he intended to "federal express" parts.

[23]A particularly perverse shortcoming in these peacetime demand histories is produced by the infrequent exercise of the purely combat systems of most tanks, ships, and planes. Actual targeting and firing was so infrequent that a peacetime history of parts could easily be much lower than the actual wartime needs. These parts, of course, tended to be the most complex and the most expensive—the most likely to be stocked at low levels in order to reduce inventory costs. They were also the most likely to involve significant lag time for production. Hence, the conditions built for a replay of a situation in which, for want of a certain chip on a certain card, a kingdom would be lost.

the friction of modern warfare.[24] In reviewing the lessons of World War II, the First Army Ordnance officer said: "Never base a plan of action on the theory that you will have enough. Base it on a probable scarcity and be ready to *manufacture* the supplies not on hand."[25]

When resupply is missing, maintenance becomes of even more critical importance, and under these circumstances it will be handicapped. Although the new doctrine depends on timely maintenance repair through the forward movement of maintenance teams, historical experience suggests these teams are not likely to emerge without enormous effort. In Vietnam, rear area maintenance units were reluctant to come forward to repair armored vehicles, forcing combat units to tow all salvageable vehicles back to the rear areas, increasing turnaround time and decreasing available combat power.

In the hypothetical example of the theater commander controlling ballistic computers parts, informal improvisation can lessen responsiveness. In attempts to speed their own planning, subordinate echelons may intercept or pool M1 parts to avoid delay in requesting spares from the theater commander. This response would increase speed and continuity in planning in subordinate echelons, but it would also undoubtedly increase the higher-level commander's problem in knowing what is needed where and when. For example, in the fall of 1944 the U.S. maintenance battalions of the 69th Ordnance Group had to move their companies forward so quickly that they did not have time to coordinate with their group headquarters. As a result, these battalion commanders were told not to move companies forward without the approval of the Corps Ordnance Officer (Mayo 1968:276–77). From then on, the speed of maintenance unit responses to battlefield conditions depended on coordination with and approval by the corps behind the division.

[24]One of the major frustrations commonly expressed by ground forces commanders during the U.S. buildup of troops in Saudi Arabia in 1990 was the unexpected slowness of sea transport of heavy weapons such as M1 tanks.

[25]See, respectively, Starry 1978:223–24 and Mayo 1968. Quotation from Mayo 1968:297 (emphasis added). In the 1973 Arab-Israeli War, the Israeli army lost nearly half (840) of its entire inventory of 2,000 tanks in a ten-day period. It survived because in that same period its maintainers were able to repair and return to action 400 of the damaged tanks (Dupuy 1978:609).

The fourth requirement, responsiveness, also directly involves speed in the sudden alteration of existing relationships to meet a changing tactical situation, but it too is likely to be difficult with the new complex organization. Heavy with parts and test equipment, logistics units cannot move as quickly as the combat units and generally cannot do their job while traveling (Reiss and Lee 1986:25). The more complex the network of support relationships, the more difficult movement becomes and the more likely it is that the friction of war will intervene. The "surging" that occurs will be limited, but when it does occur, ripple effects will move through the rest of the support system. In many respects, the M1 tank can move only as fast as its support vehicles, which carry critical, scarce parts.[26]

A hypothetical example helps describe how speed in the U.S. Army in a major theater like Europe could be affected. Assume a knowable outcome—for example, M1 ballistic computers experience a higher failure rate than expected. This failure rate occurs under four conditions: usage rates over 100 miles a day in rainy weather combined with extensive idling and use of the main gun. This problem might never have surfaced in peacetime because this combination of events rarely occurs in exercises. In addition, a shortage of repair parts occurs in the theater. The theater commander's response is to increase control of these parts at his command level, making the release of any given part more time-consuming. A logistic control of this type would require significant coordination up the chain of command to the theater level, which may not be able to respond rapidly and accurately to multiple requests from battalions desperate for parts. A constraint like this is common in war. General Patton's pursuit of the Germans in central France in 1944 ground to a halt because of a theater-level decision to cut back Patton's fuel allocation (O'Connell 1989).

With speed and accuracy impaired overall, complex tactical

---

[26]A British military expert and former officer, Richard Simkin, claims that an army needs an order-of-magnitude increase in its average speed to prevail against a stronger enemy (Wass De Czege 1986). For all of its costs and reliance on the supply system, even an unfettered M1 unit would not provide this increase over Soviet forces.

forces have difficulty in peacetime anticipating future operations, integrating operations accurately, providing continuity, being responsive, and improvising. The peacetime training must match the wartime requirements. If today's soldiers are unable to alter their operations to meet new crises, they will act as they learned to do in peacetime—and if they act wrong, die. In the Grenada invasion of 1983, despite the announced goal of rescuing students, airborne forces did not land near either campus. Trained to take airfields, they first took the airfield. Then, with all surprise lost, they marched remarkably slowly to the first rescue objective.[27] The Iraqis required most of the eight-year war with Iran in the 1980s to shake off their peacetime training with Soviet doctrine of highly centralized control of units. For the most part, their maneuvers were blunt, slow, and costly in lives and equipment (O'Ballance 1988).

What soldiers cannot do well in peacetime they also cannot do reliably in wartime, irrespective of doctrine. However, doctrine can help or hinder their wartime capabilities. Some may argue that the AirLand Battle doctrine as constructed now is the problem and that the complex machines under some other doctrine would perform effectively. However, any doctrine which uses complex machines embodying scarce knowledge will have difficulty in wartime. Because the rogue set has unknowable unknowns, the higher level of surprise cannot be fully accommodated and is both theoretically and practically potentially more destructive. The issues for those who draw up doctrine are where and how to accommodate the knowledge burden and how to plan for surprise.

The AirLand Battle doctrine correctly states that the "tempo and complexity of modern combat rule out a system that requires complicated or time-consuming coordination" (FM 100-5 1982:7-21). Yet peacetime implementation of the doctrine guided the planners to choose equipment that induced a more complex organization and less speed and accuracy in responses. As a result, a familiar cycle is likely to develop in any future large war; recognition of maintenance as a critical arena by the senior leadership will

[27]See Adkin 1989. "The tendency of troops to stick to what they have been trained for is remarkable" (Balck 1979:22).

increase the desire for control.[28] Efforts to schedule, concentrate, restrict, and monitor are also likely to expand, intensifying the intricacy of a system that would be floundering precisely because of its complexity. During the transition to war and the war itself, the cycle of positive feedback would continue: surprises induce more efforts to control, which in turn produce complexity and more surprises. The complex organization ultimately has great difficulty responding quickly to surprises.

SMALLER-SCALE OPERATIONS

A major land conflict in Europe has become much less likely in recent times, and military planning has begun to focus on smaller-scale operations and smaller armies. Depending on the equipment decisions made in reducing the size of a task force or of the entire organization, the effects of complexity in machines may not change proportionately with the changes in scope. As long as the knowledge is scarce in the wider society, the equipment will remain expensive in per unit costs. And as long as the machines are complex, they will have large rogue sets that will impose knowledge burdens on the organization. The key question is the extent to which the smallness in scale affects the ability to provide knowledge rapidly and accurately when and where it is needed.

A large-scale organization that conducts small-scale operations is more likely to have within the organization the resources to meet unexpected demands than it would have in conducting large-scale operations. Speed of response, more than accuracy, would be affected, especially in maintenance. It is common practice for units about to be tested in peacetime to borrow sufficient working equipment from fellow units to survive the exercise and then to return the equipment. If M1 tanks were needed in an operation overseas and black boxes are monitored, these boxes would be dedicated for and sent to the units most likely to see

[28]"The direct and rapid access to information tempts senior commanders to intervene in those areas for which their subordinate commanders are responsible" —"The result will be [and is] called 'battlefield management,' [which reduces] subordinate commanders to the role of executing agencies fully dependent on the level of support decided by their superiors" (Haasler and Goebel 1982:70).

combat. Other units would see an increase in the delay for deliveries of parts. Contractors can be gathered and sent; budget limits can be temporarily suspended. In the fall of 1990, Marine Corps leaders concerned that Marine equipment was not working well in the Saudi desert were considering establishing "strike teams" to fly to Saudi Arabia to diagnose and repair equipment ("Deployment" 1990:26). This option could not be considered in a major conflict. Small-scale operations are a trouble spot against which otherwise scarce resources can be massed.

The concentration of these resources, however, takes time, and complex systems would continue to surprise their users. One likely result is overcompensation for initial shortfalls: more troops, more supplies, more direct deliveries, more attention are showered on the combat forces than originally planned. The units located elsewhere who were shorted the redirected parts would continue to respond to their now heightened uncertainties with even more local adaptations but with less of a sense of urgency. This situation occurred during both the Vietnam and 1991 Gulf wars, when huge stocks of supplies were sent to the combat theater while units elsewhere faced severe shortages. Nonetheless, a large-scale organization is likely to have enough resources to provide scarce knowledge, albeit at some delay spent in gathering the material, in order to conduct a single small-scale operation. Multiple small-scale operations, however, are likely to stress the organization in ways equivalent to the large-scale conflict already discussed.

At the other end of the scale continuum, a small-scale army conducting small-scale operations would, in principle, show the same characteristics as a large-scale organization in a major conflict. Smaller armies tend to be smaller in relatively healthy Western nations because there is a political disinclination to allocate large defense budgets and hence the high costs of individual weapon systems will continue to constrain the numbers purchased, the training afforded, and the parts accumulated. Presuming that the small-scale operation stresses the organization as a major conflict would a large-scale army, the effects should mirror those already discussed. Both speed and accuracy would be affected.

The most likely outcome of force reductions in the U.S. Army, however, is a smaller-scale army responsible for both large-scale and small-scale operations. In many respects this combination is the most dangerous, because the smaller organization has fewer resources to gather for small-scale operations and proportionately less with which to hold the line while the reserves are mobilized. The urgency managers feel to reduce uncertainty is likely to be great and to exacerbate their tendency to try to create precision in responses. Stronger still will be the desire to use technical complexity to create force multipliers, enhance precision in coordination, and monitor operations to control for the unforeseen. To the extent that the larger-scale mission or multiple smaller operations are emphasized, equipment capabilities and, later, vagaries are likely to drive operational choices while rippling through the smaller-scale organization. Both speed and accuracy are at risk.

All of these variations in scale have several points in common. First, complex equipment will continue to constrain operational choices. Justifying the expense of fewer, more capable machines involves making these machines the tools of first choice to accomplish any mission, whether or not they are appropriate. Also, the cost will over time eliminate other options from the inventory. A smaller war will involve the same equipment and soldiers in smaller quantities, but the same training and expected relationships. For smaller armies, this process should occur more quickly.

Second, high costs encourage a search to ensure that each machine must have the greatest possible chance of survival. Preparations for operations lengthen as backup systems and alternatives are incorporated. To the extent that the operation planned is one of those foreseen and practiced in peacetime, the time for planning special preparations is shortened by the amount already spent during routine planning. Total preparations, however, still take as long. The time it takes the average nuclear carrier to prepare for sea operations is lengthening significantly as the internal systems of the ship become more complex (LaPorte et al. 1986).

Third, because sensitivity to uncertainty is elevated as budgets tighten, it is likely that more equipment and forces are sought than may be necessary. Grenada again provides examples. Against 600 Cuban engineers and 2,000–5,000 poorly armed Grenadian militia, the U.S. landed seven battalions, or 3,000–4,000 professional soldiers. General John Vessey, then chairman of the Joint Chiefs of

[157]

Staff, noted in his response to criticism of the operation: "During the planning period ... there was no way of determining how much of the enemy force would fight. Therefore, the U.S. forces had to be sized to meet worst case conditions." The soldiers carried so much in their 120-pound backpacks as a hedge against uncertainty that they were unable to advance very far even when unopposed. When faced with setbacks in operations in Grenada, Major General Edward Trobaugh of the 82nd Airborne called for continuous reinforcements, saying, "Keep sending battalions until I tell you to stop."[29]

With complex organizations, the minimally acceptable amounts of equipment and personnel rise for small operations as well as large.[30] Much of the new complex equipment cannot be divided up and is unique in its capabilities: half a tank is not possible, and neither is half the number of parts needed to operate that tank. The greater the support needed in general, the greater the support needed in small operations as well. For example, because the United States has no modern light tank, all operations requiring a tank, whether small or not, must take a heavy tank—an M1 or an M60—or do without.

In addition, the complexity of the national military organization has virtually assured that no large or small operation is possible without a large number of actors across functional areas from all the services. In Grenada, an existing plan could have been altered to rescue the students, but it was a more traditional operation with a single service spearheading the attack. That plan was ignored, and valuable time and accuracy was lost as a large task-force containing portions of each service's special operations force was constructed in only four days (Adkin 1989:131).[31] Furthermore, in the same operation, senior U.S. planners discovered the second student campus more than eight hours before the first wave of troops was scheduled to depart but because of the number of elements to coordinate, eight hours was not enough

[29]See, respectively, Adkin 1989:344, 222, 224.

[30]In the 1991 Gulf War, the commanding general of the allied forces originally asked for five times as many forces as he ultimately received (Dowell et al. 1991:27).

[31]This pattern of extensive joint and senior HQs was replicated in both the 1989 Panama Invasion and the 1991 Gulf War.

time to alter the planned operations (Adkin 1989:141). Consequently, had the Grenadians been interested in hostages or revenge those students would have been unprotected and easy targets.

Similarly, the assessment of minimal needs rises as an adaptation to the peacetime experience and difficulties with the machines. For example, the Iran rescue effort, was aborted when less than six helicopters remained available for the actual mission. While the remaining machines had more than enough room for all the planned occupants, the commander on the ground had been persuaded by risk assessments in Washington that six was the minimum for success. A subsequent report said the forces used were expanded out of proportion to their usefulness. The commander later said he could have accomplished the mission with two (Lanir et al. 1988:109–10). Yet the normal reliability figures for these helicopters in peacetime suggested that at least three were needed; normally about 30 percent of these machines are unable to fly.

Fourth, in the coordination of operations involving a tightly coupled group of actors, certain events acquire a pivotal position because it becomes too difficult to recoordinate a new order of activities to take into account new information. In the Grenada invasion, the senior officers delayed the entire invasion until daylight, because a Navy team was unable to put lights on the runway on time as planned. All pretense to tactical surprise was lost. Only extraordinary luck and Castro's order to the Cubans not to fire saved the landing force from losing many more soldiers (Adkin 1989).

Hence, a large or small army with complex equipment and a more complex tactical organization will find that even small operations will tend to take longer to coordinate, be larger in forces applied and expense, be less able to adapt to the unexpected and be more sensitive to surprises.[32]

[32]Owing to its exceptional circumstances, the 1991 Gulf War offers few lessons for future wars. The American force structure was sized and resourced to conduct alone one and a half wars simultaneously. Together the allies had generous material/personnel inventories, extensive time to prepare, unimpeded air/ground mobility and little enemy hindrance. With this level of redundancy, a modern military organization should be able to accommodate most rogue outcomes. These

VULNERABILITY TO SURPRISE

Despite the warnings of past military thinkers, the peacetime structure created during the 1980s appears to have added significantly to the friction of war. Posing difficulties for success in peacetime, the emerging structure is likely to have similar, if not greater, difficulty performing in wartime in a major conflict.[33] It embodies knowledge, costs and inventory constraints, and interdependencies that are not likely to disappear as battle begins.[34] The doctrine explicitly plans that the structure will operate in wartime as it does in peacetime. As Sun Tsu (Wing 1988) and Balck (1979) have observed, soldiers in wartime will try to continue the organizational arrangements of peacetime, but confusion will increase—for example, with the insertion of the in-theater depot—as planned—as a critical component of wartime parts supply or with the sudden acceptance of expedient battle damage repair. Both of these activities are not practiced in peacetime and are therefore likely to be poorly implemented under the stress of war. A mismatch between the planned structure and the new constraints is likely to develop rapidly.

The consequence of increased technical complexity has been a

conditions, however, are highly unusual and more consistent with colonial conquests than the standing-start war against a numerically superior, technologically equivalent opponent envisioned in the AirLand Battle doctrine. Future conflicts are likely to be more difficult for the emerging smaller, more technologically complex, and less redundant western militaries; success will increasingly depend on fragile logistical umbilical cords and the ability to improvise accurately and rapidly in wartime crises.

[33]Brewer argues that the complex organization is less decomposable both for containment of disruptions and for analytical understanding (Brewer 1975:179, Landau 1973:535).

[34]It is common for Army officers to claim that "all this [safety, budget, psychological, accounting, privacy act or diplomacy, etc.] garbage will go away in wartime." In fact, most of it will not. The budget and accounting restrictions now imposed on battalion commanders will be moved from the battalion to the division and the corps. Instead of the battalion commander looking at the parts budget and deciding what is affordable, that commander will be counting the number of parts the division has purchased. Safety and psychological constraints will reassert themselves brutally, and in a modern televised theater the rights of the civilians of the various participating nations will be forcibly defended by allied soldiers as well. The sad part of this misperception is that the same officers and soldiers will initially try to fight the war as though these factors did not exist; people, equipment, and opportunities will be lost until the lessons are relearned.

more complex tactical force that is less able to provide the speed of maneuver and synchronization—accuracy—of effort required by the AirLand Battle doctrine. Difficult for any organization to achieve in peacetime, the doctrine's requirements are exceptionally problematical in the wartime environment of a major theater. In the battlefield, the set of unknowns increases—not decreases—with complex equipment and complex organizations. No longer are combat service support operations the least crucial rearward functions. Their success becomes intricately involved in the viability of all other functions and operations. An organization increasingly characterized by intricate structural relations is unable to perform trial-and-error investigations accurately and rapidly—successfully.

The outcome of a purely conventional war fought from a standing start under demanding conditions, in a major theater with the AirLand Battle doctrine and the emerging complex forces is likely to be a blood-filled stalemate in which attrition, not maneuver, dominates operations. The new doctrine is often labeled a "maneuver" doctrine because of its emphasis on moving the right combat force to the right place at the right time. This emphasis on accuracy also supports the independent fighting encouraged by the manual.[35] For scarce forces to make the best of their combat assets independently, those units need to be accurately placed, but a complex organization has less ability to be both rapid and accurate. Unable to implement doctrine as it was intended, a complex military organization may be unable to avoid a war of attrition. A plan to conduct short wars of maneuver has serious consequences if the planners are wrong. By 1943 the German general staff had exactly what it had hoped to avoid: a war of attrition pitting scarce German resources against the much more abundant stocks of the Allies.

The steady pace of nuclear proliferation has increased the chances that future conventional opponents may have to moderate nuclear capabilities as well. It is likely to prove difficult to keep war between nuclear powers conventional if one side is

---

[35]See, respectively, Wass de Czege 1986, Steinbruner and Sigal 1983:100, and Cavanaugh 1983:10. "Maneuver... is the means of concentrating forces in critical areas to gain the advantages of surprise, position and momentum which enable small forces to defeat large ones" (FM 100-5 1982:7-7).

experiencing unacceptable losses. The more catastrophic initial conventional losses appear, the greater the image of irreversible failure in the perceptions of senior leaders. In these calculations of destruction and survival, an ebb could easily appear to be complete defeat (Feld 1977:72). The Army doctrine states: "It is of paramount importance that . . . nuclear release be requested in sufficient time to allow employment when he [the enemy] is still 24–60 hours from the FLOT [front lines]" (TRADOC PAM 525-25 9181:14). By the letter of this doctrine, then, the process for release would have started in the communications channels as the enemy was moving toward the front and before enemy forces had crossed any borders.[36]

An image of conventional failure can accelerate the move toward nuclear responses. A pernicious but subtle process is involved in the development of perceptions of failure. Peacetime experiences with complex equipment in a constrained organization encourage troops to distrust their equipment and the system behind them. Special efforts for successful exercises, high levels of wrongful removals, expensive parts in short supply, frequent scrounging for key material or information, and weapon systems nonfunctioning for obscure reasons do not build confidence. The problem is that troops lacking confidence give in to confusion more quickly. In the winter of 1944, the 106th Division collapsed when it was surprised by unexpected enemy attacks. Support was slow in coming, and fear fragmented unit after unit. Some 7,000–9,000 men and enormous amounts of equipment were destroyed or captured (Whiting 1981:xviii).[37] In a system of intricate support relations, the collapse of one can ripple through the system, producing the perception of collapse by all.

---

[36]It is useful to consider what might have happened if Iran had had a small nuclear arsenal when Khomeini, refusing to give up his dream of taking over Iraq, decided to allow military action against Shia populations in Iraqi cities. The Iranian forces began long-range artillery and aerial bombing of Iraqi civilian targets. The war was, by this time, a bloody stalemate, and one can only guess what Khomeini would have permitted if a more powerful weapon than "human wave" tactics had been at hand. See O'Ballance 1988.

[37]The rapidity of collapse of the Iraqi forces during the 1991 Gulf War may be another example.

[8]

The Costs of Complexity

Complexity imposes costs on an organization. For constrained organizations, the costs of a larger knowledge burden and more rogue outcomes can be especially high. To operate and survive, organizations need a minimum amount of certainty; more complexity, however, reduces organizational predictability. The least costly introduction of complexity into the organization requires an understanding of the increased knowledge needed to keep an acceptable level of certainty in the system.

Militaries face particularly hard choices. Acutely aware of their operational unknowns, they seek equipment, structures, and procedures that reduce uncertainty in combat units. Military managers lean toward the more complex machines because they offer broad capabilities and ease of use. Especially desirable have been machines that promise particularly rapid or accurate operations—those have proven difficult to resist.

The level of complexity in the new machines, however, raises the uncertainty of operations for the entire organization, from the front lines to the corps and depots. Unless accommodated, the knowledge burdens of the new weapons induce greater complexity in the using organization. The image of independent battalion or brigade, isolated and beset, fighting on its guts and wits, is false. The costs and expertise behind expensive, multilayered printed circuit boards, for example, rule the use and capabilities of new advanced weapon systems. The knowledge they entail has proven scarce; they cost more, and fewer of them are procured.

[163]

When tanks with these boards are damaged, there needs to be someone there who is capable of repairing such machines both fast and accurately because there are too few in the inventory. Because knowledge was scarce, however, that repairer is likely to be kept in the rear of the battlefield by supervisors who are unwilling to risk losing scarce parts or human expertise in front-line combat. Thus, the organization changes itself in its efforts to make the weapons work in the front-line companies.

A field army increasing in complexity has difficulty with its own rogue outcomes. When it attempts to conduct operations in wartime, it faces a large set of both knowable and unknowable unknowns and is less likely to be able to determine accurately the sources of disruptions and failures. The internal intricacy of its relationships is less likely to be able to change quickly in order to meet unforeseen requirements, especially if a clever enemy is deliberately disrupting essential operations. It will be less like the older electro-hydraulic systems that leaked, creaked, belched, and sparked when they failed, and more like the modern integrated computer—which when it fails gives few hints about what was disrupting operations; it simply stops.[1] A complex tactical organization runs the risk of doing the same.

THE TRUE COST OF THE M1

The U.S. Army embarked on its modernization with a poor understanding of the consequences of introducing complex weapon systems into its tactical organizations. Its managers attempted to compensate for the anticipated difficulties while benefiting from the promises. In formal changes, the result was a more complex tactical organization in which operations required closer coordination between maintenance and supply elements from the

---

[1]Older electro-hydraulic equipment tended to have tolerances built into the connections so that it degraded slowly and gracefully. The newer equipment is so dependent on the whole range of nodes being operational that malfunctions tend to produce a sudden halt to all operations. Furthermore, while older equipment tended to make noises or leak, announcing imminent failure in advance, the newer equipment is electronically interconnected, so that in most cases no noise or leaks give warnings of imminent failure. See Sarna 1982, Bragg and Demchak 1982, and Wohl 1980.

depot in the United States to front lines anywhere in the world.

The M1 led the bow wave of this modernization and provided the most striking example of difficulties, but other new weapons systems too imposed greater knowledge requirements on tactical organizations.[2] Together they buffeted the field organizations and encouraged ad hoc fragile arrangements that solved a temporary problem and created conditions for a fiasco in a major conflict.

Putting the M1 in the combat battalions increased the requirement that brigades, divisions, and corps be more technically knowledgeable in order to keep the level of surprise under control. It was necessary to train mechanics that specialized in the M1 and to add costly training by reintroducing theory in courses. Increases in the number of mechanics with higher scores on mental aptitude tests also were requested; skill levels, not solely the number of mechanics, became pressing issues for commanders.

As the Army's managers and commanders reacted to the higher levels of uncertainty by increasing control, complexity increased among tactical units. By intensifying efforts to concentrate, monitor, schedule, and restrict, the managers and commanders increased the complexity of the structure of the units as the complexity of the job for these units increased. The job became more demanding because the number of systems, parts, and potential maintenance problems increased. Diversity rose with the variety of systems and with the number of military specialists required to do the repairs.

Interdependence also grew. The unit-level interactions with the supply system were more closely controlled through screening and canceling procedures at higher levels. Unit authorizations to buy were restricted, unit submission of material into the system for repair or replacement invoked closer screening by technical inspectors. There was widespread use of manufacturer representatives and commodity command representatives at all levels of the theater organization; these people were valued as much for their ability to expedite the arrival of parts as for their technical expertise. Unit stockage was more strictly controlled. For many

[2]See Demchak 1987: chap. 6 app. for a review of some evidence on other systems indicating similar increases in the knowledge burden.

scarce components, access to larger theater stocks had to be specifically authorized. Approval required paperwork and the assignment of people and time to monitoring activities.

As a result, the organization evolved into groups of interconnected units whose actual capabilities are not well understood beyond the immediate vicinity of the field units involved. The near-term readiness requirements, given the technical complexity of the machines, fostered a wide variety of ad hoc arrangements that were invisible beyond the immediate array of units. As these adaptations were discovered by the cost-conscious higher head-quarters, stricter controls were imposed to eliminate the unauthorized activities. The tighter controls increased the rigidity of the system, and the correct parts and people became more difficult to obtain. The tactical units came to be more dependent on outside sources for material they could not order and stock on hand and frequently did not truly understand. Much misdiagnosed equipment flowed outward from puzzled people in the front units, linking them more tightly to the rest of the organization. This imperiled the combat units' sense of autonomy, increased their uncertainty, encouraged efforts to find informal access to missing knowledge and parts, and increased the unknowns about the entire organization, to which managers soon began to respond again. A cycle of encouraging maladaptations developed and continued.

The increase in the complexity of tactical organizations conflicted with the doctrine that was intended to bring about success in wartime. The entire organization's ability to respond quickly and accurately was weakened. Some adaptations were unit-by-unit informal agreements and very fragile. Others were made at the division level to control costs, fostering interactions that are difficult to alter when a war begins. The various arrangements will be even more difficult for replacement troops to learn and use, since each division's arrangements will be different. When a war starts, the massive introduction of a wide variety of equipment, old and new, will stress already fragile interactions. In addition, the enemy will be actively trying to disrupt these relations. Whether or not any doctrine as planned would bring success, no doctrine can be successful with an organization that conflicts with its requirements and those of wartime.

[166]

What were the lessons to be learned? First, it was not that advanced technology was always bad, but rather that certain forms of "high tech" make matters worse if the rogue sets of the machine and of the organization, respectively, remain large. Second, extensive rogue sets are endemic with complex systems, and hence, while complex systems are not harmful for all organizations, the concomitant scarcity of knowledge induces a constrained organization to become more complex and more uncertain. Therefore, two ways to help mitigate the problem are (1) to reduce scarcity of knowledge and (2) to think systemically.

Acquire Scarce Knowledge

Organizations need to recognize that scarcity of knowledge costs. If their managers purchase something built with knowledge that was scarce in wider society, such as highly complex specialized machines, and insert them into critical functions, the organization will to pay for this scarcity in time and/or resources when rogue outcomes occur. The less time and fewer resources available at the point of critical disruptions, the higher the costs of having neglected to acquire scarce knowledge or accommodate rogue outcomes. Several rules of thumb about design, testing, and change are useful in avoiding these costs.

Design Requirements. One necessary design rule is to consider not whether the design is evolutionary but how relatively scarce the knowledge needed for each design is. An old, rarely used complex machine may require knowledge more scarce than the new calculator, but anything electronic poses a special knowledge burden just because of the difficulty in observing functions. The leak in an old machine may give the observer information that a newer machine would not provide so easily. But if both are equally complex, a large rogue set in either will pose problems.

A second rule in design is to seek a low level of coupling between the machine and the many organizational elements. The farther information has to travel, the more accuracy is required at

[167]

each node and the less tolerant the system is of imprecision in diagnosis, parts, or procedure. To operate rapidly and accurately, complex systems are less able to tolerate variations in operation or maintenance. Complex systems require more knowledge, if only to assure that actions beyond the operational tolerance of the system are prevented. Technical complexity not only increases the length of time necessary to achieve such a level of familiarity, but also reduces the array of easy modifications. If the initial model is complex, then the follow-up models will also be expensive to make and will carry a heavy knowledge burden as well.[3]

Complex designs, burdened with a heavy requirement for knowledge and precision, hinder emergency battlefield modifications. Battlefield repair of airplanes with baling wire and paper departed with World War I, but the need for such expedients never disappeared. Acquiring this information is particularly expensive, because the actual circumstances under which such damage will occur are not easily reproducible in peacetime. Shooting up several fully loaded tanks only comes close.

Test Requirements. However well designed, any system will face rogue outcomes; a complex system has more of them. Extensive early testing of the completed system increases the chances of accommodating these outcomes before they ripple through the deployed units. Problems waived or ignored in pre-issue tests have consistently appeared later in the deployed units and cost more for the organization to accommodate at that time (Kaiser and White 1983, Augustine 1988:82).

The Israeli Merkava is one of the better modern tanks designed for the knowledge conditions of its army (Hellman 1985). It was designed by a small group of senior officers who were also successful veterans of modern tank combat; the tank designers emphasized their experiences by focusing on crew and ammunition battlefield survivability (Ogorkiewicz 1985:39). In particular,

---

[3]In aviation, a new model in an aircraft series usually means the maintainers have to acquire new knowledge. The machines are complex, and despite the common name the commonalities are usually tied to the airframe itself, not to the avionics inside. Hence, the maintainer of an older model might seriously disrupt the operation of the newer model if he or she does not understand the intricacies of the new relationships among subcomponents.

[168]

the process of design, test, and initial production took nine years, including two years of exhaustive testing (Foss 1986:45). The care with which this tank's design was physically tested is similar to the type of focus on durability exhibited by the Army's technical services prior to World War II, and it has produced results. For example, because of the novel engine-in-the-front internal arrangement, a Merkava crew in the 1982 Lebanon War survived two direct frontal hits from a more powerful enemy tank; in the American tanker tradition, such hits would have killed one crew member and badly injured two others (Hellman 1985:95). The designers clearly succeeded.

A large increase in the extent and validity of testing is necessary to accommodate unknowns. This alternative would take a great deal of courage, because the political costs of honest, rigorous testing occur long before any benefits are seen. The temptation to construct self-congratulatory tests that serve only to ratify decisions and justify costs is nearly overwhelming. The Merkava designers are not more pure—only more motivated toward honesty because of the ever-present threat of real combat. It requires strong incentives and herculean efforts on the part of the organization's leaders to direct the managers to construct tests that evaluate the machine realistically.

Stable Knowledge Requirements. What is known as the "learning curve" is the level of understanding about the unknowns of a system at any given point in the system's life. Each major change in the arrangement of elements in a machine or in the organization of course introduces more unknowns. The more frequent the changes, the higher the level of unknowns is likely to be. The rate of change exacerbates the accommodation of unknowns in a complex system.

Frequent changes in equipment or personnel mean knowledge will be scarcer at any given moment. New equipment will have new unknowns; new maintainers will have gaps in their knowledge. Stabilizing one or both of these adds to the level of possible accommodation of unknowns. In particular, the frequent turnover of enlisted personnel, rampant in all the American military forces, directly affects the ability of maintainers to acquire knowledge.

[169]

Stabilizing either equipment and personnel, or both, allows the accumulation of knowledge to grow; it is then easier to deal with the unexpected.

The military organizations, however, are not pausing to permit this kind of accommodation. By 1988, the Army institutional level had decided to begin yet another family of main battle tanks to be fielded in the early 1990s.[4] In addition, between the fielding of the original M1 and this newer tank, two new versions of the M1 that are substantially different from all previous versions, the M1A1 and M1A2, have been fielded. Over the period of fifteen years, then, the Army's managers plan to introduce a new tank with a new rogue set every four to five years. There is little time to move upward on the learning curve. A large rogue set is likely to be permanent, not transitional.

Learn to Think Systemically

The less the managers invest in thinking and acting systemically at the outset, the less accommodated the rogue set is and the more disastrous the consequences could be. Complexity poses a greater threat for some organizations than for others. These organizations, especially militaries, need to weigh and monitor carefully the amount of technical complexity introduced into their organizations and to select options that encourage smaller rogue sets.

Without managerial efforts to mitigate the knowledge burden, the expensive unknowns of the complex weapon system initiate a positive feedback cycle. The U.S. Army unwittingly began this process when it opted for a more complex tank whose knowledge burden was large; the scarce knowledge was likely to be expensive. In the 1970s, the Army was intent on modernization and especially on incorporating state-of-the-art electronic capabilities. In practice, affording this increase in technical sophistication required heavy investments in the initial weapon system, and fewer funds were available for ancillary requirements, such as training and parts. The Army's managers strove to resolve the

[4]See "Deployment" 1989:26. For a review of budgetary evidence, see Spinney 1985.

conundrum by buying still more technology to reduce the cost of training and parts and make the new equipment more reliable. Entailing much scarce knowledge, the equipment perversely added more uncertainty and increased perceived need for more control of the organization. And so the cycle continued, producing a more complex organization at each step.

Thinking systemically—thinking beyond each particular system to the burden on the entire organization—is essential to curbing this cycle. Senior Army leaders need to acquire information about the knowledge burden the proposed equipment puts on the *entire* organization, and then to compare the promised gain in capabilities against the systemic costs. As a rule of thumb, the more "state of the art" that is incorporated into the machine, the more the unknowable unknowns are present, the greater the scarcity of knowledge, and ultimately the greater the organizational costs of that machine.

AN ARSENAL OF KNOWLEDGE?

Tailoring the knowledge burden so that rogue outcomes are more fully accommodated given the circumstances of the organization is essential. This means reducing the scarcity of knowledge *before* the machine is introduced in the organization and dampening the ripple effects. In essence, the organization is systemically comparing promised equipment benefits with organizational costs. One alternative is to alter the organization's structure. What may be needed is a new arsenal concept that can reduce scarce knowledge within the organization itself while taking advantage of technical advances in the wider society.

A three-part division of military technical innovation has developed over the years: arsenal (in-house), industrial (useful but not directly military), and laboratory (unrelated basic research that is later purely military). The last two types of innovation are done by civilians outside of the organization and later adapted to military purposes. Currently the Army's managers deal with all three types, but the emphasis is on the latter two for systems with electronics (Feld 1977).

[171]

In particular, the basic design of critical weapon systems is created by civilians whose decisions largely determine the knowledge burden of the machine. The difficulty is that knowledge which seems common to the design engineers may be exceptionally scarce in the society at large and the Army in particular. Also, key decisions about the level of coupling internally in the machine is left in the hands of the individuals least likely to be available to the organization later when the machine enters the deployed forces. The Army acquires the machine whose knowledge burden is set by well-educated specialists external to the organization.

One solution is to avoid scarce knowledge, often called "buying off the shelf"; however, this alternative is part of the problem. Off-the-shelf equipment that has been highly coupled in the final machine can present an enormous knowledge burden as well. Furthermore, the tendency to buy state-of-the-art, whether or not it is "off the shelf," means that several rogue sets can act synergistically to create an overall rogue set greater than the sum of the individual sets. Hence, instructing civilian companies to buy off the shelf is only a partial answer to the problem; it does not address the integration of the subcomponents tailored by an understanding of the knowledge conditions of the organization itself. Although it is in this group of decisions that the rogue set is initially constructed; reliable subsystems can combine to produce a much less reliable final system (McNaugher 1989:101).

The Merkava example may provide an alternative: an arsenal in which skilled members of the organization performed this vital design integration using innovations available in the wider society. A key function of this arsenal would be the design and extensive testing of different arrangements of subsystems to determine the interactive effects before a final system is produced. This concept has several advantages. First, it leaves the innovation and production of basic electronics in the private sector, where it is better served. Once the private sector disseminates its products so that the knowledge is less scarce, the Army's technically trained soldiers and officers could pick up various subsystems and integrate them into an Army machine.

Second, this idea moves the decisions about integrating the rogue sets of the subsystems into the hands of the members of

[172]

the organization most likely to know how rapid and accurate knowledge needs to be in the deployed units.[5] Merged into this testing could be the views of combat veterans.[6] Janowitz observed that military veterans can develop sufficient prestige to "checkmate the advice of the scientist and technologist" (Janowitz 1971:24).

The goal is to limit the effect of the technological bias and tie the products systemically to the wartime reality of rogue outcomes and a lack of knowledge. Along with an emphasis on building and testing complete systems would be a strong interest in decoupling the organizational elements around new equipment. Because new machines that cannot operate or be repaired without a large support structure are more likely to induce tighter coupling in the organization, this kind of machine should be avoided. Implementing this arsenal concept would make the knowledge about the machine intrinsically less scarce at its first deployment than it would be coming whole from private contractors. It would also make the effects of the machine's knowledge burden less surprising in the organization.

An arsenal of this type would initially be expensive and run the historically frequent risk of being slow to implement innovations. Acceptance of a longer cycle in production, however, that means more rogue outcomes are likely to be identified before the frontline units face the equipment. The "knowledge arsenal" would have the advantages of favoring products with obvious military benefits, incorporating the influence of combat (if available) or exercise experiences, and—very important—retaining Army proprietary control of plans and parts designs.[7] Civilian innovations would still be available for purchase and integration into Army

[5]In order to have the experience of service in the military, these arsenal employees, if not on active duty, should at least be required to serve in the reserves. There is legal precedent in the Army's and Air Force's current reserve forces.

[6]Given the age and currency of most American tank combat veterans, these would probably have to be Israelis. The last major tank combat against a robust opponent for the United States occurred in World War II. See my comments in Chapter 7 on the lessons of the 1991 Gulf War.

[7]In the early 1980s, the Army established its "High Technology Test Bed" battalion at Fort Lewis, Washington. It was intended to test off-the-shelf, relatively exotic equipment for possible military use, but unfortunately it became associated with images of soldiers dashing about on dune buggies rather than with images of serious testing. It was dismantled in the late 1980s owing to cost.

arsenal test programs. Fuller reportedly said that soldiers are alchemists (Holley 1953:15), and, if true, this tendency would have advantages in the arsenal concept. There the soldiers could mix and test innovative combinations of established private developments to maximize the available knowledge and minimize the burden.

Army officers could become innovators without being deviants (Feld 1977:88). An understanding of rogue outcomes is more likely to travel through the organization as officers slowly cycle in and out of the arsenal. The organization's managers will be more sensitive to the need to dampen the ripple effects of the unavoidable unknowable unknowns. The cycling of officers at long intervals encourages changes in the organization's ideology that will produce more knowledgeable senior officers who reinforce incentives for rigorous testing and smaller rogue sets. It also allows direct feedback from the deployed units.[8] In this structural alternative, the organization would have stronger internal control of the knowledge burdens of machine that will be critical to future battles.

THE QUALITY VERSUS QUANTITY DEBATE

The concept of a knowledge arsenal conflicts with both sides of the quality versus quantity debate that raged in the defense community during the 1970s and 1980s. Although such an arsenal would rely on civilian innovators for new subsystems, it would not be incorporating the state-of-the-art materiel that the quality advocates would prefer, and while it would be testing extensively, it would not be as cheap as the quantity promoters prescribe. In addition, neither side would be pleased with how long it took to design and test a new system. Those on the two sides of the

---

[8]A lieutenant-colonel who had participated directly in constructing the military essential needs statement (MENS) for the DIVAD air defense weapon was perplexed by the failure of the program. He said he could not explain how the Army's simple desire to have a gun that shot faster and farther resulted in the DIVAD (Private conversation, 1984). It is important to note that, once the MENS was finished, construction of the DIVAD was then left in the hands of civilian contractors, not military officers.

debate, however, have been talking past each other for twenty years, and it is in that conceptual gap that the problems of complex equipment need to be addressed. The scarcity of knowledge makes complex systems troublesome and surprising, but both sides have not been able to take this into account systemically.

The argument for quality is based on a presumption that the technological superiority of a particular weapon or set of weapons will overcome the numerical superiority of an attacking enemy. In its crudest form, this premise leads to the conclusion that the more numerically superior the enemy, the more necessary it is to have the technological quality to meet the threat. Therefore, technical advancements are powerful "force multipliers" and should be incorporated into the force as soon as possible. The focus is on extending and elaborating the capabilities of individual weapon systems as soon as possible or affordable (Perry 1984, Coates and Killian 1985:358).

This approach ignores the knowledge burden imposed on the entire system while heralding the ease of operating or the enhanced capabilities of a single weapon system. Also forgotten are organizational problems in general and the bounds of rationality that make organizations so essential (Williamson 1975). The Army has taken this approach over the last twenty years and has also paid for scarce knowledge in large organizational costs. The quality side of the argument tends to encourage the technological imperative and a predilection to large, complex machines with many capabilities. It is a reductionist approach, as opposed to a systemic approach, and it leaves the scarcity of knowledge as an externality to which the organization has had to adapt.

The quantity argument recognizes the difficulties presented by complexity and a lack of information. As its name implies, however, the preferred solution is many more weapons. To get a multitude of weapons under budgets, they would have to be cheap—but inexpensive weapons are possible only if the knowledge entailed in the machine is not scarce. Hence, numerous cheap and simple weapons are proposed to meet the threat of a numerically superior enemy (Spinney 1985). In short, redundancy is the solution to complexity and the knowledge burden.

But the problem is that the understanding of complexity is not

[175]

sufficiently developed.[9] One indicator of complexity is the number of components. If the numbers increase without a commensurate decrease along the other indicators, then complexity and its knowledge burden are not decreased. Great numbers of simple, cheap weapons actively in use also can mean great numbers of coordinating elements in the organization—and thus an increase in interdependence which is not automatically accommodated by the machines. For example, in World War II the sheer density of heavy bombers limited their usefulness because of the possibility of fratricide (Van Creveld 1989:194).

On both sides of the debate, the knowledge burden on the organization should be the key discriminator: the more scarce the necessary knowledge when a rogue outcome occurs, the less acceptable that machine should be. The burden is greater if the machine is complex because the level of unknowns—especially unknowable unknowns—is higher irrespective of the quantity or quality.

Rogue outcomes are universally present and should be treated accordingly. Other organizations, such as the Air Force or the Navy, with long histories of familiarity with complex machines, have shown only slightly lower rates of false removals of machine components despite years of experience. Even the Army's own technical suborganization, the missile units using the HAWK missile system for twenty-odd years, units that never suffered from the decline in aptitude as the others, tend to exhibit high false-removal rates (Nauta 1983). This common characteristic both supports the ubiquity of rogue outcomes and suggests that, in a complex system, the accommodations necessary to make the rogue set smaller are not easy to make.

On the other hand, if knowledge is not scarce in the society or organization, then it is easier to accommodate the rogue outcomes. The organization is more likely to be able to buy knowl-

---

[9]This is not to equate all military reformers with this characterization of the debate. One reformer, John Boyd, a retired Air Force colonel and combat ace, has worked tirelessly in the defense community to promote the idea of focusing on or avoiding surprise in battle rather than numbers or quality of machines (Coates and Killian 1985:268).

edge, perhaps even extra machines. For example, Army radios are electronic and internally somewhat complex, but the knowledge is not scarce. Radios are common in the society at large. In the organization, both radios and parts are relatively abundant, and because radios have a long history of use, knowable unknowns are well-known. Therefore, training is less expensive and repair is in principle less problematic.[10] This equipment does not tend to invoke the heightened sense of uncertainty that induces the control mechanisms of the managers and the associated ripple effects. If the knowledge in the M1 were as common as that of the radio, the organization could buy enough M1s to stock extras to accommodate even some unknowable unknowns. Having a multitude of rogue outcomes is no longer as important if the organization can throw away the broken machine and use a new one.

Finally, both sides (and others) tend to blame the long lead time in the development of weapon systems for the later costs and problems of the system. Usually the solution proposed is to streamline the process, and the past successes of wartime improvisations are used as support for this alternative. However, making the process quicker will not necessarily improve the product or eliminate surprises. Wartime improvisations occur in the light of real-time, real-fire testing on the battlefield; they also tend to be incremental rather than radical. Without increased attention to and testing for rogue outcomes and its systemic interactions during design and development, a quicker delivery in peacetime means only that the tactical forces will face surprises sooner with less knowledge.

A Look to the Future of Our Case Study

The future U.S. Army will face the same problems of the current Army; it will be perpetually in transition trying to meet

---

[10]Army radios do differ to some extent from civilian radios because of the security devices that scramble and unscramble transmissions. These additional capabilities elevate the scarcity of knowledge and the costs of the otherwise technically common devices.

the advances of the enemy. Its structure, however, is likely to show signs of responding more to the precision requirements of the new machines than to the needs of the Army's traditional weapon, the soldier. Seemingly unbeknown to most of the major actors, the rush for advanced technology began a fundamental change in the organization. In the organization that traditionally placed personnel over equipment in decisions, now equipment needs drive planning decisions.

The Army of the future is more likely to resemble the Air Force, an organization in which a large number of technically differentiated and highly interdependent individuals operate and support a relatively small number of weapons. The trends in the Army toward greater technical differentiation and interdependence show no sign of stopping as of yet. In the current era of a lessening of tensions in the European theater, the major Western armies are planning on becoming smaller and relying heavily on advanced complex weapons for the advantage in deterrence or war. Barring some resurgence of a major conventional threat, future armies are likely to be smaller in numbers of soldiers but possibly no less complex, with increases in differentiation and interdependence compensating for fewer numbers.[11] A smaller army may not mean an army that is more deployable or effective. If its equipment keeps it intricately dependent on a fragile network of support, the smaller army will show the same problems as the larger army—but with less firepower.

These changes are not immutable, however, and senior Army leaders can choose to make modernization more thoughtful in its selections by tailoring and targeting the knowledge burden. This effort will generally entail avoiding state-of-the-art technology and complexity in the most critical systems, and it will involve seeking out the knowable unknowns by extensive testing in integrating diverse subsystems and using the knowledge that members of the organization have in the design of this integration. The knowledge arsenal concept could provide many of these machine cor-

[11]The internal pressures toward greater numbers of people are stymied in the active force by personnel ceilings. If this imposed restraint is lifted, the natural pressures to expand would certainly prevail.

rections and help with general education of the organization. The rest is up to the senior leaders and their responses to the costs of complexity. The costs of doing nothing could not be greater.

A Step in the Right Direction

This book has both practical and theoretical value. Understanding the reasons behind the high costs of complex technical machines, and knowing how to measure complexity, will make it possible for managers to understand the character of the uncertainty contained in these machines. And an appreciation of the inevitability of rogue outcomes can help managers make decisions about how much knowledge to seek, how much uncertainty to accept, where to place the knowledge obtained, and perhaps most important, how much effort will be necessary to substitute for unobtainable knowledge—that is, how much accommodation will cost.

The organizational changes described in this book reinforce the practical lesson, that "to manage is not to control" (Landau 1979). There is not necessarily a connection between a manager's control and the knowledge required to reduce uncertainty. In the Army, efforts to control variations in maintenance so as to reduce the knowledge required in spare parts resulted in greater interdependence and greater rigidity in the organization with respect to crises. The trade-off is a local reduction in uncertainty that precipitated an increase in uncertainty for the rest of the tactical forces. The local units saw no other choice but to reduce uncertainty, and this local adaptation is part of the lesson for managers who are uncritically enthusiastic about technical advances.

The acquisition and potential advantages of these complex machines are to be approached cautiously and thoughtfully. Public organizations and their audiences need to be particularly attentive to the implications of complex machines. Such organizations are often charged with tasks in areas where the means to perform and the costs involved are not completely known, and where the uncertainty with regard to successful performance is

[179]

high. Organizations limited by societal injunctions from making mistakes will find that decisions surrounding the introduction of complex technical machines are especially perilous. Savings in money, time, or labor may not occur,[12] and the uncertainty within the organization can rise, rather than fall as intended. The concepts and applications suggested here can help managers in these organizations think more thoughtfully about the technical choices they make.

This book also contributes to the central issues of organization theory: how to explain the functioning of, changes to, and learning of organizations. The problem is that organizations as systems are inherently composed of complex intricate relationships that challenge human thinking processes. If humans think in relatively linear patterns—sequentially and limited in the number of factors that can be accurately and simultaneously manipulated—then some way to make a theoretical leap beyond the limitations of the structure of the human mind is required in order to understand complexity. To understand complexity, human beings must find ways to comprehend complex interactions without thinking in intricate patterns—perhaps by summarizing them to chunks or branches amenable to our thought patterns. The unexpected nonlinear aspects of human thinking are poorly understood (Dreyfus), but it is there that the complete conceptual understanding of complexity may need to be developed. In many ways, a variety of disciplines are attempting to make this leap in their fields of expertise.

This book draws on the work of mathematicians and engineers as well as sociologists and organization theorists. The concept of complexity as captured in the rogue set or a subsequent formulation offers one way to cross academic discipline lines, to benefit from advances in other disciplines, and to make more likely the theoretical leap to understanding complex relations in organizations, machines, and other systems. Critical complex organiza-

---

[12]The university librarian of the University of California observed: "With very few exceptions, the introduction of automation has not meant tangible dollar savings to total library costs. . . . What it has meant is that we do things . . . faster. . . . We have not laid off anybody because of automation . . . [it] has helped reduce the amount of tedious work" (Rosenthal 1984:17).

tions in our society are relying more and more on complex technical machines, with the result that organizational changes that can have powerful effects are developing virtually unnoticed. Organization theory needs a mechanism for identifying these changes *before* they occur and for evaluating the possible organizational responses once they have occurred. This work is a necessary step in that direction.

Bibliography

Adkin, Mark, 1989. *Urgent Fury: The Battle for Grenada*. Lexington, Mass.: Lexington Books.

AFSC (Armed Forces Staff College). 1983. "Joint Staff Officer's Guide." AFSC Publication 1. Washington, D.C.: National Defense University.

ALC (Army Logistics Center). 1978. "Proposed Revision to Career Management Field 63 (Mechanical Maintenance)." Washington, D.C.: Department of the Army.

Alford, Jonathan. 1983. "Perspectives on Strategy." In *Alliance Security: NATO and the No-First-Use Question,* edited by John D. Steinbruner and Leon V. Sigal, 91–102. Washington, D.C.: The Brookings Institution.

——. 1984. "Review of Conventional Deterrence by John Mearsheimer." *Bulletin of Atomic Scientists*, May, 41–42.

Allison, Graham T. 1971. *Essence of Decision*. 1st ed. Boston: Little, Brown.

ALOG. 1980. "CMF 63 Changed." *Army Logistician* (Department of the Army), July–August, 22–27.

——. 1984a. "Logistics Hotlines Offer Expert Assistance." *Army Logistician*, November–December, 44–45.

——. 1984b. "Maintenance Diagnosticians Proposed." *Army Logistician*, March–April, 40.

——. 1984c. "Supply Career Changes Approved." *Army Logistician*, September–October, 38.

——. 1985a. "Critical Components to Get Centralized Management." *Army Logistician*, November–December. 40.

——. 1985b. "Material Quick-Alert Teams Formed." *Army Logistician*, May–June, 41.

——. 1986a. "CALS Planned." *Army Logistician*, November–December, 1.

——. 1986b. "Maintenance Policy Changes Set." *Army Logistician*, November–December, 37.

——. 1986c. "More Officer Logisticians Needed." *Army Logistician*, November–December, 42.

Anton, Thomas J. 1980. *Administered Politics*. Boston: M. Nijhoff.

AR 750-1. 1978. "Army Material Maintenance Concepts and Policies" (Army Regulation). Washington, D.C.: Department of the Army.

———. 1983. "Army Material Maintenance Concepts and Policies" (Army Regulation). Washington, D.C.: Department of the Army.

ARD&AM. 1985. "Evolution of M60 Tank Series Continues." *Army Research, Development & Acquisition Magazine*, March–April, 28.

Armed Forces Journal. 1984. "Industry Forecasts Reinforce Inman's View of the Future." *Armed Forces Journal International*, September 87–91.

"Armor School Hotline Now Open." 1984. *Armor* (Department of the Army), July–August, 50.

Armoured Vehicles. 1981. London: Jane's Publishing Company.

Armstrong, John A. 1973. *The European Administrative Elite*. Princeton: Princeton University Press.

Army Times. 1986. "Wagner Assesses Weapons Modernization." *Army Times*, 17 October, 27–28.

Art, Robert J. 1968. *The TFX Decision: McNamara and the Military*. Boston: Little, Brown.

Augustine, Norman R. 1986. (1983). *Augustine's Laws*. New York: Viking Press.

AUSA. 1986. "Command and Staff." *Army Magazine* 36 (October), 284–352. In *Greenbook* (annual publication). Washington, D.C.: Association of United States Army.

"Aviation Logistical Issues." 1986. *Aviation Digest* (Department of the Army) 32 (November), 28.

AvWk. 1991. "Program Offers Rapid Solutions to Problems Encountered by F-15Es Engaged in Gulf War." *Aviation Week and Space Technology*, 4 February, 59.

Baker, Caleb. 1991. "Army Aircraft Need Better Care, Report Says." *Army Times*, 12 November, 27.

Balck, Hermann. 1979. "Translations of Taped Conversations with General Hermann Balck." Manuscript, Battelle Tactical Technology Center, Columbus, Ohio.

Bamford, James. 1982. *The Puzzle Palace*. Boston: Houghton Mifflin.

Barnaby, Frank. 1986. *The Automated Battlefield*. New York: The Free Press.

Barnard, Chester I. 1968. (1938). *The Functions of the Executive*. Cambridge: Harvard University Press.

Barnett, Corelli. 1986. *The Collapse of British Power*. Atlantic Highlands, N.J.: Humanities Press.

Barth, Gary, ed. 1978. *Scale and Social Organization*. Oslo, Norway: Universitetsforlaget.

Beard, Edmund. 1976. *Developing the ICBM: A Study in Bureaucratic Politics*. New York: Columbia University Press.

Bergerson, Frederic A. 1980. *The Army Gets an Air Force: Tactics of Insurgent Bureaucratic Politics*. Baltimore: Johns Hopkins University Press.

Bigleman, Paul A. 1981. "Division and Corps 86: Force Designs for the Future." *Army Magazine* 31 (June), 23–33. In *Greenbook* (annual publication). Washington, D.C.: Association of United States Army.

Binkin, Martin. 1986. *Military Technology and Defense Manpower*. Washington, D.C.: The Brookings Institution.

Blair, Bruce G. 1985. *Strategic Command and Control*. Washington, D.C.: The Brookings Institution.

Blau, Peter M., Cecilia McHugh Falbe, William McKinley, and Phelps K. Tracy. 1976. "Technology and Organization in Manufacturing." *Administrative Science Quarterly* 21 (March), 1–15.

Bond, David F. 1991. "Apache Helicopter Proves Reliability to Rebut Reputation for Deficiencies." *Aviation Week and Space Technology*, 4 March, 25.

Booth, William. 1990. "Investigating a Billion-Dollar Blunder." *Washington Post Weekly*, 16–22 July, p. 32.

Brachman, Raymond. 1968. "Economics of Introducing Automatic Diagnostic Equipment to Organizational Maintenance (ATE/ICEPM-ORGL)." Memo Report M68-32-2. Frankford Arsenal: Department of the Army.

Bracken, Paul. 1983. *The Command and Control of Nuclear Forces*. New Haven: Yale University Press.

Bragg, Lucas J., and Chris C. Demchak. 1981. "An Approach to Evaluating Maintenance Difficulty." Draft report no. ML 101. Washington, D.C.: Logistics Management Institute.

——. 1982. "An Approach to Evaluating Maintenance Difficulty." Report no. ML 101. Washington, D.C.: Logistics Management Institute.

Bramson, Leon, and George W. Goethals, eds. 1968. *War: Studies from Psychology, Sociology, and Anthropology*. New York: Basic Books.

Brewer, Gary. 1975. "Analysis of Complex Systems: An Experiment and Its Implications for Policy Making." In *Organized Social Complexity*, edited by Todd R. LaPorte, 175–219. Princeton: Princeton University Press.

Brinberg, David, and Joseph E. McGrath. 1985. *Validity and the Research Process*. Newbury Park, Calif.: Sage Publications.

Brodie, Bernard, and Fawn Brodie. 1973. *From Crossbow to H-Bomb: The Evolution of the Weapons and Tactics of Warfare*. Bloomington: Indiana University Press.

Brown, Frederic J. 1985. "Manning Issues Revisited." *Armor Magazine*, March–April, 6–7.

Butler, Walter. 1982. "Training Developments Study—Bradley Fighting Vehicle Unit—Conduct of Fire (UCOFT) Trainer." Report no. TRASANA-TEA-28-82. Fort Monroe, Va.: TRADOC Systems Analysis Activity.

Bykofsky, Joseph, and Harold Larson. 1957. *The Transportation Corps: Operations Overseas*. U.S. Army in World War II. Washington, D.C.: Department of the Army.

Caplow, Theodore. 1964. *Principles of Organization*. New York: Harcourt, Brace.

Carney, Larry. 1985a. "Sixty-two New Courses Planned to Equalize NCO Training." *Army Times*, 25 March, 39.

——. 1985b. "Squeeze on Army Training Space Increases in Germany." *Army Times*, 15 July, 18.

——. 1986a. "Master Gunners to Be Trained for Bradleys." *Army Times*, 3 February, 17.

——. 1986b. "NCOs Sought for Technical Fields." *Army Times*, 10 March, 34.

Casti, John L. 1984. "Systems Complexity." *Options: Journal of the International Institute of Applied Systems Analysis* 3:6–9.

——. 1989. *Alternate Realities: Mathematical Models of Nature and Man*. New York: John Wiley & Sons.

Cavanaugh, Charles G. Jr. 1983. "Airland Battle." *Soldier Magazine*, July, 6–11.

CBO (Congressional Budget Office). 1983. "Cost Growth in Weapon Systems: Recent Experience and Possible Remedies." Report prepared for Senate

Government Affairs Committee. Washington, D.C.: Congressional Budget Office.

Chandler, Alfred D. Jr. 1962. *Strategy and Structure*. Cambridge: M.I.T. Press.

———. 1977. *The Visible Hand: The Managerial Revolution in American Business*. Cambridge: Harvard University Press.

Churchman, C. West. 1968. *The Systems Approach*. New York: Dell.

Clark, Asa A. IV, et al., eds. 1984. *The Defense Reform Debate*. Baltimore: Johns Hopkins University Press.

Clausewitz, Carl von. 1976. *On War*. Edited and translated by Michael Howard and Peter Paret. Princeton: Princeton University Press.

Coakley, Robert W., and Robert M. Leighton. 1968. *Global Logistics and Strategy, 1943–1945*. Washington, D.C.: War Department, Historical Division.

Coates, James and Michael Kilian. 1985. *Heavy Losses: The Dangerous Decline of American Defense*. New York: Viking Press.

Cohen, Michael D., James G. March, and Johan P. Olsen. 1972. "A Garbage Can Model of Organizational Choice." *Administrative Science Quarterly* 17 (March), 1–25.

Craig, Gordon A. 1955. *The Politics of the Prussian Army, 1640–1945*. London: Oxford University Press.

Crow, Duncan, and Robert J. Icks. 1975. *Encyclopedia of Tanks*. Secaucus, N.J.: Chartwell Books.

Crozier, Michel. 1964. *The Bureaucratic Phenomenon*. Chicago: University of Chicago Press.

Cummings, Arthur L. 1980. "Comparison Test of Tank Combat, Full-tracked M60 (4000 Miles)." Aberdeen Proving Grounds. Aberdeen, Md.: Department of the Army.

Cushen, Edward. 1982. Research memo from visit to Army Program Manager Office for the M1 Abrams tank. Logistics Management Institute, Washington, D.C.

Cyert, Richard, and James G. March. 1963. *A Behavioral Theory of the Firm*. Englewood, N.J.: Prentice-Hall.

Daft, Richard L., and Norman B. Macintosh. 1981. "A Tentative Exploration into the Amount and Equivocality of Information Processing in Organizational Work Units." *Administrative Science Quarterly* 26 (June), 207–24.

Dahl, Robert A., and Edward R. Tufte. 1972. *Size and Democracy*. Stanford, Calif.: Stanford University Press.

DA PAM 310-1. 1983. *Consolidated Index of Army Publications and Blank Forms*. Washington, D.C.: Department of the Army.

DA PAM 360-866. 1982. "Force Modernization." *Commander's Call* (Department of the Army), January–February.

DA PAM 750-35. 1983. "Functional User's Guide for Motor Pool Operations." Washington, D.C.: Department of the Army.

DARCOM. 1981. "This Is United States Army Material Development Command." Pamphlet, U.S. Army Material Development Command. Washington, D.C.: Department of the Army.

Davies, W. J.K. 1984. (1973). *German Army Handbook, 1939–1945*. New York: Arco.

Deane, John R. 1976. "United States Army Material Development Command." *Army Magazine*, October, 78–89. In *Greenbook* (annual publication). Washington, D.C.: Association of United States Army.

Demchak, Chris C. 1981. Interviews with U.S. Army maintenance training staff at Aberdeen Proving Grounds, Maryland.

——. 1983. Interviews with personnel, U.S. Army units in Europe and continental United States.

——. 1984. Interviews with members of Congress and staff on military authorization or appropriation committees. Washington, D.C.

——. 1987. *War, Technological Complexity, and the United States Army*. Ph.D. dissertation, University of California at Berkeley.

——. 1988. Interviews with German scholars and members of *Bundeswehr* units and agencies, Federal Republic of Germany.

Department of Defense. 1981. *Soviet Military Power*. Washington, D.C.: Department of Defense.

Department of the Army. 1984. (1950). "Rear Area Security in Russia: The Soviet Second Front Behind German Lines." Pamphlet 20-240. Washington, D.C.: U.S. Army Center of Military History.

"Deployment May Reignite Army-Marine Feud." Defense Trends. 1990. *Army Times*, 10 September, 26.

Dewar, Robert, and Jerald Hage. 1978. "Size, Technology, Complexity, and Structural Differentiation: Toward a Theoretical Synthesis." *Administrative Science Quarterly* 23 (March), 111–36.

Dickson, Paul. 1971. *Think Tanks*. New York: Atheneum.

DIV Ref. 1979. *Division Reference Data*. ST-7-1-1. Washington, D.C.: Department of the Army.

DOD. 1990. "Manpower Requirements Report—FY 1990." Department of Defense. Report submitted to the Congress of the United States.

Donnelly, Tom. 1985. "Realism Sought in New Live-fire Tests of Bradley." *Army Times*, 17 June, 24–25.

——. 1986. "Army's Bradley Program Dodges Bullet in House Vote." *Army Times*, 1 September, 27.

Dowell, William, Bruce van Voorst, and Robert T. Zintl. 1991. "The 100 Hours." *Newsweek*, 11 March, 22–32.

Downs, Anthony. 1967. *Inside Bureaucracy*. Boston: Little, Brown.

Drucker, Peter F. 1977. *Technology, Management, and Society*. New York: Harper & Row.

Dunnigan, James F. 1982. *How to Make War: All the World's Weapons, Armed Forces, and Tactics*. New York: William Morrow.

DuPicq, Charles Ardant. 1946. *Battle Studies: Ancient and Modern Battle*. Translated by John N. Greely and Robert C. Cotton. Harrisburg, Pa.: Military Service Publishing Co.

Dupuy, Trevor N. 1978. *Elusive Victory: The Arab-Israeli Wars, 1947–1974*. New York: Harper & Row.

——. 1980. *The Evolution of Weapons and Warfare*. New York: Jane's Publishing Group.

Durkheim, Emile. 1947. *The Division of Labor in Society*. 1st ed. Glencoe, Ill.: The Free Press.

Earle, Edward Meade, ed. 1971. (1943). *Makers of Modern Strategy*. 1st ed. Princeton: Princeton University Press.

Easterbrook, Greg. 1984. "Why the DIVAD Wouldn't Die." *Atlantic Monthly*, November, 10–22.

Ellul, Jacques. 1964. *The Technological Society*. New York: Vintage Books.

Engel, John D. 1968. "A Revised Job Requirements Inventory for General Vehicle Repairmen MOS 63C." HUMMRO Study ES-66. DTIC no. AD701608. Washington, D.C.: Department of the Army.

English, John A. 1984. *On Infantry*. New York: Praeger.

Erlich, Paul, Ann Erlich, and John P. Holdren. 1977. *Ecoscience*. San Francisco: W. H. Freeman.

Etzioni, Amitai, ed. 1969. *The Semi-Professions and Their Organization*. New York: The Free Press.

Fallows, James. 1981. *National Defense*. New York: Random House.

Feld, Maury D. 1977. *The Structure of Violence: Armed Forces as Social Systems*. Inter-University Seminar on Armed Forces and Society. Beverly Hills, Calif.: Sage Publications.

Felsenthal, Dan S. 1980. "Applying the Redundancy Concept to Administrative Organizations." *Public Administration Review* 40 (May–June), 247–52.

Fiddle, Seymour, ed. 1980. *Uncertainty: Behavior and Social Dimensions*. New York: Praeger.

Fitts, P., and M. Posner. 1967. *Human Performance*. Belmont, Calif.: Brooks/Cole.

FM 21-6. 1943. *List of Publications for Training*. Army Field Manual. Washington, D.C.: War Department, (Department of the Army).

FM 100-5. 1976. *Operations*. Army Field Manual. Washington, D.C.: Department of the Army.

———. 1982. *Operations*. Army Field Manual. Washington, D.C.: Department of the Army.

———. 1986. *Operations*. Army Field Manual. Washington, D.C.: Department of the Army.

FM 100-16. 1983. *Support Operations: Echelons Above Corps*. Draft Army Field Manual. Washington, D.C.: Department of the Army.

"Forces in Place in the Gulf." 1991. *New York Times International*, 25 February.

Foss, Christopher F. 1986. *Jane's Main Battle Tanks*. 2nd ed. London: Jane's Publishing Group.

Fox, Ronald J. 1974. *Arming America: How the United States Buys Weapons*. 1st ed. Cambridge: Harvard University Press.

———. 1984. *Arming America: How the United States Buys Weapons*. 2nd ed. Cambridge: Harvard University Press.

Fulghum, David. 1986. "Supercomputer Will Help Army Design Its Weapons." *Army Times*, 22 September, 38–39.

Gabriel, Richard A. 1985. *Military Incompetence*. New York: Hill & Wang.

Gabriel, Richard A., and Paul L. Savage. 1978. *Crisis in Command*. New York: Hill & Wang.

Galbraith, Jay R. 1977. *Organizational Design*. Reading, Mass.: Addison-Wesley.

GAO (General Accounting Office). 1981a. "Logistics Planning for the M1 Tank: Implications for Reduced Readiness and Increased Support Costs." Report no. PLRD-81-33. Washington, D.C.

——. 1981b. "NORAD's Missile Warning System: What Went Wrong?" Report no. MASAD-84-30. Washington, D.C.

——. 1981c. "The Services Should Improve Their Processes for Determining Requirements for Supplies and Spare Parts." Report no. PLRD-82-12. Washington, D.C.

——. 1982. "Improving the Effectiveness and Acquisition Management of Selected Weapon Systems: A Summary of Major Issues and Recommended Actions." Washington, D.C.

——. 1983a. "The Army Should Confirm Sergeant York Air Defense Gun's Reliability and Maintainability before Exercising Next Production Option." Report no. MASAD-82-8. Washington, D.C.

——. 1983b. "Poor Procurement Practices Resulted in Unnecessary Costs in Procuring M1 Tank Spares." Report no. PLRD-83-21. Washington, D.C.

Gerth, H. H., and C. Wright Mills, trans. and eds. 1978 (1946). *From Max Weber: Essays in Sociology*. 1st ed. New York: Oxford University Press.

Gessert, Robert A. 1983. "The Airland Battle and NATO's New Doctrinal Debate." Paper delivered at the 21st Annual Conference of the Council on Christian Approaches to Defense and Disarmament. Oegstgeest, Netherlands.

Gibson, James William. 1986. *The Perfect War*. New York: Vintage Books.

Gividen, George M. 1981. "1981 Apportionment Review: Assessment of New Systems Briefing for Office of Under Secretary of Defense, Research and Engineering." Washington, D.C.: Army Research Institute.

Goodwin, Jacob. 1985. *Brotherhood of Arms: General Dynamics and the Business of Defending America*. New York: Times Books.

Green, Constance M., Harry C. Thomson, and Peter C. Roots. 1955. *The Ordnance Department: Planning the Munitions for War*. Washington, D.C.: Department of the Army.

Greenfield, Kent R., Palmer Robert R, and Bell I. Wiley. 1947. *The Army Ground Forces: The Organization of Ground Combat Troops*. United States Army in World War II. Washington, D.C.: Department of the Army.

Gregg, Arthur J. 1980. "Logistics as a Force Multiplier." *Army Magazine*, October, 134–36. In *Greenbook* (annual publication). Washington, D.C.: Association of United States Army.

Griffith, Robert K., Jr. 1982. *Men Wanted for the United States Army*. Westport, Conn.: Greenwood Press.

Guthrie, John R. 1977. "United States Material Command." *Army Magazine*, October, 56–67. In *Greenbook* (annual publication). Washington, D.C.: Association of United States Army.

——. 1979. "DARCOM: Old Solutions Will No Longer Suffice." *Army Magazine*, October, 50–55. In *Greenbook* (annual publication). Washington, D.C.: Association of United States Army.

——. 1980. "Not a Change in Mission, but a Change in Intensity." *Army Magazine*, October, 54–63. In *Greenbook* (annual publication). Washington, D.C.: Association of United States Army.

Haasler, Ruprecht, and Hans Goebel. 1982. (1981). "Uneasiness about Technical Progress in the Armed Forces." *Military Review*, October, 62–72.

HAC (House Appropriations Committee). 1971. "Committee Report Accompanying H.R. 11731, DOD Appropriations Bill FY72." 92nd Congress, 1st session.

——. 1971a. "Hearings, DOD Appropriations for FY 72, Part 5." 92nd Congress, 1st session.

——. 1971b. "Hearings, DOD Appropriations for FY 72, Part 1." 92nd Congress, 1st session.

——. 1975. "Hearings, Subcommittee on Defense." 94th Congress, 1st session.

——. 1981. "Hearings, DOD Appropriations Bill FY 82." 97th Congress, 1st session.

Halperin, Morton H. 1974. *Bureaucratic Politics and Foreign Policy.* 1st ed. Washington, D.C.: The Brookings Institution.

Hampton, David R. 1978. *Behavioral Concepts in Management.* 3rd ed. Belmont, Calif.: Wadsworth.

Handel, Michael I., ed. 1986. *Clausewitz and Modern Strategy.* Totowa, N.J.: Frank Cass.

Hanks, Chris H. 1981. "The Effects of Increased Transportation Times on Spares Funding Requirements and Aircraft Availability Rates." Report for Task AF 101. Washington, D.C.: Logistics Management Institute.

HASC (House Committee on Armed Services). 1985. "Hearings for DOD Authorization for FY 85." 99th Congress, 1st session.

Hastings, Max, and Simon Jenkins. 1983. *The Battle for the Falklands.* New York: W. W. Norton.

Hellman, Peter. 1985. "Israel's Chariot of Fire." *Atlantic Monthly,* March, 81–95.

Hewes, James E. Jr. 1975. *From Root to McNamara: Army Organization and Administration, 1900–1963.* Washington, D.C.: U.S. Army Center of Military History.

Hilmes, Rolf. 1987. *Main Battle Tanks: Developments in Design Since 1945.* Translated by Richard Simpkin. London: Brassey.

Hittle, J. D. 1952. (1947). *Jomini and His Summary of the Art of War.* Harrisburg, Pa.: Military Service Publishing Company.

Hochmuth, Milton S. 1974. *Organizing the Transnational.* Leiden: A. W. Sitjhoff.

Hoffman, Martin R. 1976. "Secretary of the Army." *Army Magazine,* October, 8–11. In *Greenbook* (annual publication). Washington, D.C.: Association of United States Army.

Hofstadter, Douglas R. 1980. *Godel, Escher, Bach: An Eternal Golden Braid.* New York: Vintage Books.

Holley, I. B. 1983. (1953). *Ideas and Weapons: Exploitation of the Aerial Weapon by the United States During World War I, A Study in the Relationship of Technological Advance, Military Doctrine, and the Development of Weapons.* Washington, D.C.: Office of Air Force History.

Hoos, Ida R. 1983. (1972). *Systems Analysis in Public Policy: A Critique.* 2nd ed. Berkeley: University of California Press.

Hough, Jerry F. 1969. *The Soviet Prefects: The Local Party Organs in Industrial Decision-making.* Cambridge: Harvard University Press.

House of Representatives. 1983. "Conference Report on the DOD Authorization Act of 1984 to Accompany S675." 98th Congress, 1st session.

Hunnicutt, R. P. 1978. *Sherman: A History of the American Medium Battle Tank.* Novato, Calif.: Presidio Press.

——. 1984. *Patton: A History of the American Main Battle Tank.* Novato, Calif.: Presidio Press.

Huston, James A. 1966. *The Sinews of War, 1775–1953.* Army Historical Series. Washington, D.C.: Department of the Army.

[189]

Huzar, Elias. 1971. (1950). *The Purse and the Sword: Control of the Army by Congress through Military Appropriations, 1933–1950*. Westport, Conn.: Greenwood Press.

International Defense Review. 1989. "Israeli Defense Plans Reviewed." *International Defense Review* 22 (June 1), 723.

Janis, Irving L., and Leon Mann. 1971. *Decision-making*. New York: The Free Press.

Janowitz, Morris. 1964. *The New Military: Changing Patterns of Organization*. New York: Russell Sage Foundation.

———. 1971. *The Professional Soldier*. 1st ed. New York: The Free Press.

Jomini, Baron de. 1971. *The Art of War*. Translated by G. H. Mendell and W. P. Churchill. Westport, Conn.: Greenwood Press.

Kahn, Alfred E. 1968. "The Tyranny of Small Decisions: Market Failures, Imperfections, and Limits of Econometrics." In *Economic Theories of International Politics*, edited by Bruce M. Russett., 52. Chicago: Markham.

Kaiser, Robert D., and Richard M. Fabbro. 1980. "DOD Use of Civilian Technicians." Report no. ML 004. Washington, D.C.: Logistics Management Institute.

Kaiser, Robert D., and Thomas A. White. 1983. "Improving Weapon System Support." Report No. ML 210. Washington, D.C.: Logistics Management Institute.

Kaldor, Mary. 1981. *The Baroque Arsenal*. London: Andre Deutsch.

Kane, John J. 1981. "Personnel and Training Subsystem Integration in an Armor System." Report no. SAI-81-332-WA. Washington, D.C.: Science Application Incorporated.

Kantrow, Alan K. 1980. "The Strategy-Technology Connection." *Harvard Business Review* 58 (July–August), 6–21.

Kaprellian, Caspar. 1985. "Failure-Free Organizations." Seminar delivered at the University of California at Berkeley.

Kaufman, Herbert. 1971. *The Limits of Organizational Change*. University: University of Alabama Press.

———. 1981. *The Administrative Behavior of Federal Bureau Chiefs*. Washington, D.C.: The Brookings Institution.

Kaufman, William W. 1983. "Nonnuclear Deterrence." In *Alliance Security: NATO and the No-First-Use Question*, edited by John D. Steinbruner and Leon V. Sigal. Washington, D.C.: The Brookings Institution.

Keegan, John. 1977. *The Face of Battle: A Study of Agincourt, Waterloo, and the Somme*. New York: Vintage Books.

Keith, Donald R. 1982. "United States Army Material Development Command." *Army Magazine* 32 (October), 58–68. In *Greenbook* (annual publication). Washington, D.C.: Association of United States Army.

King, Edward L. 1972. *The Death of the Army*. New York: Saturday Review Press.

Kjellstrom, John A. 1976. "Managing the Army's Money, '76." *Army Magazine* 26 (October), 91–97. In *Greenbook* (annual publication). Washington, D.C.: Association of United States Army.

Kleinman, Forrest K., and Robert S. Horowitz. 1964. *The Modern United States Army*. New York: Van Nostrand.

Kosakevic, Victor. 1982. Interview. Washington, D.C.

Kraemer, Kenneth L., and James L. Perry. 1989. "Innovation and Computing in the Public Sector: A Review of the Research." *Knowledge and Society: The International Journal of Knowledge Transfer* 2(1):72–87.

Kraus, Ken E. 1986. "Leader's Fiscal Responsibility." *Army Logistician* 18 (November–December), 27.

Lambright, W. Henry. 1976. *Governing Science and Technology.* New York: Oxford University Press.

Landau, Martin. 1969. "Redundancy, Rationality, and the Problems of Duplication and Overlap." *Public Administration Review* 29 (July–August), 346–58.

——. 1973. "On the Concept of a Self-correcting Organization." *Public Administration Review* 33 (November–December 1973), 533–42.

Landau, Martin, and Russell Stout Jr. 1979. "To Manage Is Not to Control: The Folly of Type II Errors." *Public Administration Review* 39 (March–April), 148–56.

Landes, David S. 1969. *The Unbound Prometheus: Technological Change and Industrial Development in Western Europe from 1750 to the Present.* Cambridge: Cambridge University Press.

Lanir, Zvi, Baruch Fischoff, and Stephen Johnson. 1988. "Military Risk-taking: C3I and the Cognitive Functions of Boldness in War." *Journal of Strategic Studies* 11 (March), 96–114.

LaPorte, Todd R., ed. 1975. *Organized Social Complexity.* Princeton: Princeton University Press.

——. 1987. "High Reliability Organizations: The Problems and Dimension of the Research Challenge." Unpublished paper. University of California at Berkeley.

LaPorte, Todd R., Karlene Roberts, and Gene I. Rochlin. 1986. "Current Research on Failure-free Organizations." Unpublished paper. University of California at Berkeley, Institute of Governmental Studies.

Layton, Edwin T. 1985. *And I Was There: Pearl Harbor and Midway, Breaking the Secrets.* New York: William Morrow.

LEA (Logistics Evaluation Agency). 1979. "XM1 Tank ILS Program: Interim Assessment Report." Washington, D.C.: Department of the Army.

Leavitt, Harold J., and Louis R. Pondy, eds. 1973. (1964). *Readings in Managerial Psychology.* 2nd ed. Chicago: University of Chicago Press.

Lincoln, James R., and Jon Miller. 1979. "Work and Friendship Ties in Organizations: A Comparative Analysis of Relational Networks." *Administrative Science Quarterly* 24 (June), 181–99.

Lindblom, Charles E. 1980. *The Policy-making Process.* 2nd ed. Englewood Cliffs, N.J.: Prentice-Hall.

Lussier, Frances M. 1986. *Army Air Defense for Forward Areas: Strategies and Costs.* Congressional Budget Office report no. 61-3310-86-1 Washington, D.C.: U.S. Congress.

Luttwak, Edward N. 1984. *The Pentagon and the Art of War.* New York: Simon & Schuster.

McCurdy, Howard E. 1977. *Public Administration: A Synthesis.* Menlo Park: Cummings.

Machiavelli, Niccolo. 1952. *The Prince.* Translated by Luigi Ricci, edited by E. R. P. Vincent. New York: New American Library.

——. 1970. *The Discourses.* Translated by Leslie J. Walker, revised by Brian Richardson. Middlesex, U.K.: Penguin Books.

Macksey, Kenneth, and John H. Batchelor. 1970. *Tank: A History of the Armoured Fighting Vehicle.* New York: Charles Scribner's Sons.

McNaugher, Thomas L. 1981. "Collaboration Development of Main Battle Tanks: Lessons from U.S.-German Experience." RAND Note N-1680-RC.

——. 1989. *New Weapons, Old Politics: America's Military Procurement Muddle.* Washington, D.C.: The Brookings Institution.

McNeill, William H. 1982. *The Pursuit of Power: Technology, Armed Force, and Society since A.D. 1000*. Chicago: University of Chicago Press.

McPherson, James M. 1988. *Battle Cry of Freedom: The Civil War Era*. New York: Oxford University Press.

Maier, Charles S. 1975. *Recasting Bourgeois Europe*. Princeton: Princeton University Press.

Maitland, A. J., et al. 1981. "Effectiveness of Training—OT/III M1 (Abrams) Main Battle Tank." TRASANA-TEA-37-81, vols. 1–4 Washington, D.C.: TRADOC Systems Analysis Activity, Department of the Army.

Mansfield, Edwin. 1971. *Technological Change*. New York: W. W. Norton.

March, James G., ed. 1965. *Handbook of Organizations*. New York: Rand McNally.

———. 1988. *Decisions and Organizations*. New York: Basil Blackwell.

March, James G., and Z. Shapira. 1989. "Managerial Perspectives on Risk and Risk-taking." In *Decisions and Organizations*, edited by James G. March., 76–97. New York: Basil Blackwell.

Margolis, Howard. 1987. *Patterns, Thinking, and Cognition: A Theory of Judgment*. Chicago: University of Chicago Press.

Marsh, John. 1982. "Secretary of the Army." *Army Magazine*, October, 10–14. In *Greenbook* (annual publication). Washington, D.C.: Association of United States Army.

Marshall, S.L.A. 1978. (1947). *Men against Fire: The Problem of Battle Command in Future War*. Gloucester, Mass.: Peter Smith.

Matheny, Michael R. 1985. "In the Beginning . . ." *Armor*, July–August, 28–32.

Mayhew, David. 1974. *Congress: The Electoral Connection*. New Haven: Yale University Press.

Mayo, Lida. 1968. *The Ordnance Department: On Beachhead and Battlefront*. United States Army in World War II. Washington, D.C.: Department of the Army.

Merton, Robert K. 1968. *Social Theory and Social Structure*. New York: The Free Press.

Meyer, Deborah G. 1983. "Army Rolls out Apache Two Months Ahead of Schedule." *Armed Forces Journal International*, November.

Meyer, Edward C. 1982. "Time of Transition." *Army Magazine*, October, 18–24. In *Greenbook* (annual publication). Washington, D.C.: Association of United States Army.

———. 1984. *Extracts of Public Record June 79–June 83*. Berkeley: University of California Press.

Middleton, Drew. 1983. *Crossroads of Modern Warfare*. Garden City, N.Y.: Doubleday.

———. 1985. "Will the Tank Survive." *New York Times Magazine*, 24 November, 104–12.

Millett, Allan R., and Williamson Murray. 1989. (1987). *Military Effectiveness*, Vol. 3: *The Second War World*. Cambridge: Unwin Hyman.

Mitzberg, Henry. 1979. *The Structuring of Organizations*. Englewood Clifs, N.J.: Prentice-Hall.

Moodie, Michael. 1989. *The Dreadful Fury: Advanced Military Technology and the Atlantic Alliance*. Washington, D.C.: Praeger.

Moore, Molly. 1986. "If Only Weapons Were as Simple as Soldiers." *Washington Post Weekly Edition*, 27 October, 31.

Morgan, Gareth. 1986. *Images of Organization*. Beverly Hills, Calif.: Sage Publications.

Morone, Joseph G., and Edward J. Woodhouse. 1986. *Averting Catastrophe: Strategies for Regulating Risky Technologies*. Berkeley: University of California Press.

Morrisette, Walt. 1986. "Master Diagnostician Course to Speed Repair of Vehicles." *Army Times*, 3 February, 34.

MTOE 17-035HE101. 19981. "Table of Organization and Equipment, USAREUR Armor Battalion." Washington, D.C.: Department of the Army.

MTOE 17-235E101. 1981. "Table of Organization and Equipment, USAREUR 'J' Series Armor Battalion." Washington, D.C.: Department of the Army.

Mueller, John. 1989. *Retreat from Doomsday: The Obsolescence of Major War*. New York: Basic Books.

Mueller-Hillebrand, Burkhart H. 1954. *German Tank Maintenance in World War II*. DA pamphlet 20-202. Washington, D.C.: Department of the Army.

Musashi, Miyamoto. 1974. *A Book of Five Rings: The Classic Guide to Strategy*. Translated by Victor Harris. Woodstock, N.Y.: Overlook Press.

Nagel, Ernest, and James R. Newman. 1958. *Godel's Proof*. New York: New York University Press.

Narragon, Eugene A., Chris C. Demchak, Thomas A. White, and Joseph R. Wilk. 1984. "Re-orienting Field-level Maintenance." Report no. ML 312. Washington, D.C.: Logistics Management Institute.

Nauta, Franz. 1982a. "AH-64 Automatic Test Equipment Requirements." Working Note No. ML 213-1. Washington, D.C.: Logistics Management Institute.

——. 1982b. "Fix-Forward: Comparison of the Army's Requirements and Capabilities for Forward Support Maintenance." Draft Working Note No. ML 104. Washington, D.C.: Logistics Management Institute.

——. 1983. "Fix-Forward: Comparison of the Army's Requirements and Capabilities for Forward Support Maintenance." Final Working Note No. ML 104. Washington, D.C.: Logistics Management Institute.

Nielsen, Gerald. 1983. "An Assessment of Selected Army Technical Manuals." DARCOM Technical Report No. 383. Washington, D.C.: Department of the Army.

O'Ballance, Edgar. 1988. *The Gulf War*. New York: Brassey's Defence Publishers.

O'Connell, Robert L. 1989. *Of Arms and Men: A History of War, Weapons, and Aggression*. New York: Oxford University Press.

Ogorkiewicz, Richard M. 1985. "Master Diagnostician Concept." *Armor*, November–December, 39–44.

Ogul, Morris. 1976. *Congress Oversees the Bureaucracy*. Pittsburgh: University of Pittsburgh Press.

Olson, Howard C., L. Mathers Boyd, Norman Willard Jr., and Norman E. Willmouth. 1955. "Technical Supplement to the Report on a Survey of Armor Training Problems." Washington, D.C.: Human Resources Research Office.

Olson, Mancur. 1971. *The Logic of Collective Action*. Cambridge: Harvard University Press.

Ordnance Magazine. 1986. "Master Diagnostician Concept." *Ordnance Magazine*, Winter, 37.

Ostovich, Rudolph IV. 1986. "Airland Battle." *Aviation Digest* 32 (November), 3–9.

Oxford University. 1981. (1971). *The Compact Edition of the Oxford English Dictionary*. Oxford: Oxford University Press.

Pacey, Arnold. 1978. (1974). *The Maze of Ingenuity: Ideas and Idealism in the Development of Technology*. 2nd ed. Cambridge: M.I.T. Press.

Palmer, Bruce Jr. 1984. *The 25-Year War: America's Military Role in Vietnam*. New York: Simon & Schuster.

Perrow, Charles. 1979. *Complex Organizations.* 2nd ed. Glenview, Ill.: Scott Foresman.
———. 1984. *Normal Accidents.* New York: Basic Books.
Perry, William. 1984. "Defense Reform and the Quantity-Quality Quandary." In *The Defense Reform Debate,* edited by Asa A. Clark et al., 182–91. Baltimore: Johns Hopkins University Press.
Peters, Thomas J., and Robert H. Waterman Jr. 1982. *In Search of Excellence: Lessons from America's Best-run Companies.* New York: Harper & Row.
Pierce, John R. 1980. (1961). *An Introduction to Information Theory: Symbols, Signals, and Noise.* New York: Dover.
Pizer, Vernon. 1967. *The United States Army.* New York: Praeger.
Polsby, Nelson W. 1984. *Political Innovation in America.* New Haven: Yale University Press.
Posen, Barry. 1984. *The Sources of Military Doctrine.* Ithaca: Cornell University Press.
The Power of the Pentagon. 1972. Washington, D.C.: Congressional Quarterly.
Pressman, Jeffrey, and Aaron B. Wildavsky. 1973. *Implementation.* Berkeley and Los Angeles: University of California Press.
Pugh, D. S., ed. 1971. *Organization Theory.* Middlesex, U.K.: Penguin Books.
Ranson, Stewart, Robert Hinnings, and Royston Greenwood. 1980. "The Structuring of Organizational Structures." *Administrative Science Quarterly,* 25:1–16.
Rasor, Dina. ed. 1983. *More Bucks, Less Bang: How the Pentagon Buys Ineffective Weapons.* Washington, D.C.: Fund for Constitutional Government.
Record, Jeffrey. 1981. *NATO's Theater Nuclear Force Modernization Programs: The Real Issues.* Cambridge, Mass: Institute for Foreign Policy Analysis.
———. 1988. *Beyond Military Reform: American Defense Dilemmas.* Washington, D.C.: Pergamon-Brassey.
Reich, Robert B. 1973. *The Next American Frontier.* New York: Times Books.
Reiss, William R., and Gary R. Lee. 1986. "How to Support Deep Operations." *Army Logistician* 18 (December), 22–26.
Richardson, William R. 1982. "Deputy Chief of Staff for Operations and Plans." *Army Magazine* 32 (October), 252. In *Greenbook* (annual publication). Washington, D.C.: Association of United States Army.
Rickey, Don, Jr. 1963. *Forty Miles a Day on Beans and Hay.* Norman: University of Oklahoma Press.
Rivilin, Alice. 1983. "CBO Review of Operating and Support Costs for Army Modernization." Statement before Senate Appropriations Committee, Sucommittee on Defense. Washington, D.C.: Congressional Budget Office.
Rochlin, Gene I. 1980. Personal communication. University of California at Berkeley.
Rogers, Bernard W. 1977. "Chief of Staff of the Army." *Army Magazine,* October, 22–26. In *Greenbook* (annual publication). Washington, D.C.: Association of United States Army.
Rosen, Stephen P. 1988. "New Ways of War: Understanding Military Innovation." *International Security* 13 (Summer), 134–168.
Rosenberg, Hans. 1958. *Bureaucracy, Aristocracy, and Autocracy: The Prussian Experience, 1660–1815.* Boston: Beacon Press.
Rosenthal, Joe. 1984. "Checking at the Library." *California Monthly* 94 (March–April), 17.
Rue, Donald S., and R. O. Lorenz. 1983. "Study of the Causes of Unnecessary Removals of Avionics Equipment." Report no. F30602-79-C-0200. Hughes Aircraft Company.

Runciman, W. G., ed. 1978. *Max Weber*. Cambridge: Cambridge University Press.

Sapolsky, Harvey M. 1972. *The Polaris System Development*. Cambridge: Harvard University Press.

Sarna, Donald S. 1982. "The Future of Diagnostics." *Armor*, March–April, 15–19.

SASC (Senate Armed Service Committee). 1983. "Organization, Structure, and Decision-making Procedures of DOD, parts 1-12." 98th Congress, 1st session.

SCGO (Senate Committee on Government Operations). 1976. "Hearing on Major Systems Acquisition Reform." 94th Congress, 2nd session.

Schreier, Fred. 1977. "A Tank Designed to Cost: The United States Army's XM1." *International Defense Review*, March, 459–68.

Schurman, Franz. 1968. *Ideology and Organization in Communist China*. 2nd ed. Berkeley and Los Angeles: University of California Press.

Scott, Robert. 1981. *Organizations: Rational, Natural, and Open Systems*. Englewood Cliffs, N.J.: Prentice-Hall.

Scribner, Barry L., Alton D. Smith, Robert H. Baldwin, and Robert L. Phillips. 1986. "Are Smart Tankers Better?" *Armed Forces and Society* 12(2):193–206.

Seidman, Harold. 1975. *Politics, Position, and Power.* 2nd ed. New York: Oxford University Press.

Selznick, Harold. 1949. *TVA and the Grassroots: A Study in the Sociology of Formal Organizations*. Berkeley and Los Angeles: University of California Press.

Senate Government Affairs Committee. 1982. "Report on Federal Agency Acquisition Management." 97th Congress, 2nd session.

SESAME. 1979. Data Sample of 63 Line Replaceable Units (LRUs). Washington, D.C.: Logistics Management Institute.

Shafritz, Jay M., and Philip H. Whitbeck, eds. 1978. *Classics of Organization Theory*. 1st ed. Oak Park, Ill.: Moore.

———. 1987. *Classics of Organization Theory.* 2nd ed. Chicago: Dorsey Press.

Shields, Joyce L., Stephen L. Goldberg, and J. Douglas Dressel. 1979. "Retention of Basic Soldiering Skills." Army Research Institute. Washington, D.C.: Department of the Army.

Sigal, Leon V. 1984. *Nuclear Forces in Europe*. Washington, D.C.: The Brookings Institution.

Simms, Edward D., Chris C. Demchak, and Joseph P. Wilk. 1982. "Marine Corps Reserve and Logistics Support Functions." Working Note No. ML 206-4. Washington, D.C.: Logistics Management Institute.

Simon, Herbert A. 1969. *The Sciences of the Artificial*. Cambridge: M.I.T. Press.

———. 1976. *Administrative Behavior.* 3rd ed. New York: The Free Press.

Skolnikoff, Eugene B. 1972. *The International Imperatives of Technology*. Berkeley and Los Angeles: University of California Press.

Slackman, Joel, ed. 1986. "Quality Soldiers: The Costs of Manning the Active Army." Washington, D.C.: Congressional Budget Office.

Slatkin, Nora, comp. 1982. "Army Ground Combat Modernization for the 1980s: Potential Costs and Effects for NATO." Washington, D.C.: Congressional Budget Office.

Smith, Homer D. 1986. "NATO Maintenance and Supply Agency (NAMSA)." *Army Logistician* 18 (November–December), 2–6.

Smith, Jack W. 1975. 'DSARC Tradeoff Decisions for Aircraft." Working Note No. 7501-6. Washington, D.C.: Logistics Management Institute.

Smith, Robert P. 1980. "Division 86 Logistics." *Army Logistician*, November–December, 2–5.

Spinney, Franklin C. 1985. *Defense Facts of Life: The Plans/Reality Mismatch*. Boulder, Colo.: Westview Press.

Sprey, Pierre. 1984. "The Case for Better and Cheaper Weapons." In *The Defense Reform Debate*, edited by Asa A. Clark et al., 193–207. Baltimore: Johns Hopkins University Press.

Starry, Donn A. 1978. *Mounted Combat in Vietnam*. United States Army in Vietnam. Washington, D.C.: Department of the Army.

———. 1980. "Values, Not Score, the Best Measure of Soldier Quality." *Army Magazine*, October, 38–43. In *Greenbook* (annual publication). Washington, D.C.: Association of United States Army.

Staw, Barry M., and Gerald R. Salancik. 1977. *New Directions in Organizational Behavior*. Chicago: St. Clair Press.

Steele, Lowell. 1983. "Managers Misconceptions about Technology." *Harvard Business Review* 61 (November–December), 133–40.

Steinbruner, John D. 1974. *The Cybernetic Theory of Behavior*. Princeton: Princeton University Press.

Steinbruner, John D., and Leon V. Sigal, eds. 1983. *Alliance Security: NATO and the No-First-Use Question*. Washington, D.C.: The Brookings Institution.

Steinhart, Carol E., and John S. Steinhart. 1974. *Energy*. North Scituate, Mass.: Duxbury Press.

Strauch, Ralph E. 1975. "'Squishy' Problems and Quantitative Methods." *Policy Sciences* 6(2):175–84.

Suleiman, Ezra N. 1974. *Politics, Power, and Bureaucracy in France: The Administrative Elite*. Princeton: Princeton University Press.

Summers, Harry G., Jr. 1981. *On Strategy: The Vietnam War in Context*. Carlisle Barracks, Pa.: United States Army War College.

Sun Tzu. 1963. *The Art of War*. Translated by Samuel B. Griffith. London: Oxford University Press.

Teich, Albert H., ed. 1977. *Technology and Man's Future*. 2nd ed. New York: St. Martin's.

Tetlock, Philip E. 1979. "Accountability and Complexity of Thought." *Journal of Personality and Social Psychology* 45 (November), 74–83.

Thompson, James D. 1967. *Organizations in Action*. New York: McGraw-Hill.

Thompson, Richard H. 1985. "AMC's New Emphasis: Improving Product Quality." *Army Logistician*, May–June, 2–3.

Thucydides. 1951. *The Peloponnesian War*. Translated with an introduction by Rex Warner. New York: Modern Library.

Tice, Jim. 1985. "Apache Maintenance Training Will Be Intense." *Army Times*, 21 October, 55.

———. 1986. "Army Gets Go-ahead in Air Defense Systems." *Army Times*, 1 September, 27.

———. 1987. "Army Will Cut 550 from Headquarters Staff." *Army Times*, 16 March, 31–33.

———. 1988. "Rumble of Too Many Tanks Stilled in Europe by Poor Mechanic Skills." *Army Times*, 19 September, 6.

Tichy, N., and C. Fombrun. 1979. "Network Analysis in Organizational Settings." *Human Relations* 32 (November), 923–65.

TM38-L1711. 1984. "Technical Manual, Standard Property Book System." Washington, D.C.: Department of the Army.

[196]

Tobias, Sheila, et al. 1982. *What Kinds of Guns Are They Buying for Your Butter?* New York: William Morrow.

TOE 07-245600. 1981. "Table of Organization and Equipment, Maintenance Company, Brigade Support Battalion." Washington, D.C.: Department of the Army.

TOE 07-246S600. 1981. "Table of Organization and Equipment, Heavy Maintenance Company, Armor Division." Washington, D.C.: Department of the Army.

TOE 17-235A120. 1981. "Table of Organization and Equipment, Tank Battalion Equipped M1 Abrams Tank." Washington, D.C.: Department of the Army.

TOE 17-235J (Div 86). 1982. "Table of Organization and Equipment, Tank Battalion." Washington, D.C.: Department of the Army.

TOE 17-35H. 1983. "Table of Organization and Equipment, Tank Battalion 105mm." Washington, D.C.: Department of the Army.

TOE 29-38MO. 1974. "Table of Organization and Equipment, Heavy Maintenance Company, Armor Division." Washington, D.C.: Department of the Army.

Toomepuu, Juri. 1981. "Soldier Capability: Army Combat Effectiveness (SCACE)." Report no. ADB062417. Fort Monroe, Va.: U.S. Army Soldier Support Center.

TRADOC PAM 525-2. 1980. "Tactical Command and Control." Army Operational Concepts Series. U.S. Army Training and Doctrine Command. Washington, D.C.: Department of the Army.

———. 1981. "The Airland Battle and Corps 86." Army Operational Concepts Series. U.S. Army Training and Doctrine Command. Washington, D.C.: Department of the Army.

TRADOC PAM 525-4. 1980. "Heavy Division Operations: United States Army Operational Concept." Army Operational Concepts Series. U.S. Army Training and Doctrine Command. Washington, D.C.: Department of the Army.

TRADOC PAM 525-12. 1981. "Operational Concepts for the Communications Zone Logistic Operations." Army Operational Concepts Series. Training and Doctrine Command. Washington, D.C.: Department of the Army.

Uhle-Wetter, Franz. 1981. *Battlefield Central Europe.* 3rd ed. Munich: Bernard & Graefe.

UPI. 1991. "Army Gives Weapons Good Grade, Cites Some Problems." *United Press International Wire Service*, 14 March.

Ursul, A.D. 1975. "The Problem of Objectivity." In *Entropy and Information in Science and Philosophy*, edited by Libor Kubat and Jiri Zeman, 187–200. New York: Elsevier.

U.S. Army. 1982. Unclassified charts comparing the current division structure with the Division 86 structure. Washington, D.C.: Department of the Army.

Utley, Robert M. 1973. *Frontier Regulars: The United States Army and the Indian, 1866–1891.* New York: Macmillan.

Van Creveld, Martin. 1977. *Supplying War: Logistics from Wallerstein to Patton.* Cambridge: Cambridge University Press.

———. 1982. *Fighting Power: German and United States Army Performance, 1939–1945.* Westport, Conn.: Greenwood Press.

———. 1985. *Command in War.* Cambridge: Harvard University Press.

———. 1989. *Technology in War.* New York: The Free Press.

Vanderveen, Bart H. 1982. *A Sourcebook of Military Support Vehicles.* London: Ward Lock.

Waldo, Dwight. 1980. *The Enterprise of Public Administration: A Summary View.* Novato, Calif.: Chandler & Sharp.

Walker, Arthur H., and Jay W. Lorsch. 1968. "Organizational Choice: Product Versus Function." *Harvard Business Review*, November–December, 129–38.

Walsh, Annmarie Hauck. 1978. *The Public's Business: The Politics and Practices of Government Corporations*. Cambridge: M.I.T. Press.

Waltz, Kenneth N. 1979. *Theory of International Politics*. Reading, Mass.: Addison-Wesley.

Wamsley, Gary L., and Mayer N. Zald. 1973. *The Political Economy of Public Organizations*. Lexington, Mass.: D. C. Heath.

"War at a Glance." 1991. *Arizona Daily Star*, 4 February, p. A2.

Wass de Czege, Huba. 1986. Interview (senior Army infantry officer). Fort Ord, Calif. May.

Webster, Richard D. 1982. "Attacking Logistics Problems through Acquisition Reform." *Defense Management Journal*, 4th quarter, 3–8.

Weick, Karl E. 1979. (1969). *The Social Psychology of Organizing*. 2nd ed. Reading, Mass.: Addison-Wesley.

Weigley, Russell F. 1967. *History of the United States Army*. 1st ed. Bloomington: Indiana University Press.

——. 1973. *The American Way of War: A History of United States Military Strategy and Policy*. New York: Macmillan.

——. 1984. *History of the United States Army*. 2nd ed. Bloomington: Indiana University Press.

White, Lynn, Jr. 1978. (1962). *Medieval Technology and Social Change*. Oxford: Oxford University Press.

Whiting, Charles. 1981. *Death of a Division*. New York: Stein & Day.

Wilensky, Harold L. 1967. *Organizational Intelligence*. New York: Basic Books.

Williamson, Oliver E. 1975. *Markets and Hierarchies: Analysis and Antitrust Implications, a Study in the Economics of Internal Organization*. New York: The Free Press.

Wilson, Bennie J., III, ed. 1985. *The Guard and Reserve in the Total Force*. Washington, D.C.: National Defense University Press.

Wilson, Harlan. 1975. "Complexity in Political Theory." In *Organized Social Complexity*, edited by Todd R. LaPorte, 281–331. Princeton: Princeton University Press.

Wing, R. L., trans. 1988. *The Art of Strategy: A New Translation of Sun Tzu's Classic "The Art of War."* New York: Doubleday.

Winner, Langdon. 1975. "Complexity and the Limits of Human Understanding." In *Organized Social Complexity*, edited by Todd R. LaPorte, 40–76 Princeton: Princeton University Press.

——. 1977. *Autonomous Technology*. Cambridge: M.I.T. Press.

Wohl, Joseph G. 1980. 'Diagnostic Behavior, Systems Complexity, and Repair Time: A Predictive Theory." MITRE note M80-0008. Bedford, Mass.: MITRE.

Woll, Peter. 1977. *American Bureaucracy*. 2nd ed. New York: W. W. Norton.

Yin, Robert K. 1984. *Case Study Research: Design and Methods*. Newbury Park, Calif.: Sage Publications.

York, Herbert F., and G. Allen Greb. 1977. "Military Research and Development: A Postwar History." *Bulletin of Atomic Scientists* 33 (January).

Zeman, Jiri. 1975. "Information, Knowledge, and Time." In *Entropy and Information in Science and Philosophy*, edited by Libor Kubat and Jiri Zeman, 245–54. New York: Elsevier.

Zuboff, Shoshana. 1988. *In the Age of the Smart Machine*. New York: Basic Books.

Index

Library of Congress Cataloging-in-Publication Data

Demchak, Chris C.
  Military organizations, complex machines : modernization in the U.S.
armed services / Chris C. Demchak.
      p.   cm. — (Cornell studies in security affairs)
  Includes bibliographical references and index.
  ISBN 0-8014-2468-2 (alk. paper)
      1. United States—Armed Forces—Management. I. Title. II. Series.
UB23.D44   1991
355'.00973—dc20   90-55731

The repeating of hypothesis is irritating